Nathan Lupton

The elementary principles of scientific agriculture

Nathan Lupton

The elementary principles of scientific agriculture

ISBN/EAN: 9783337278380

Printed in Europe, USA, Canada, Australia, Japan

Cover: Foto ©berggeist007 / pixelio.de

More available books at **www.hansebooks.com**

THE

ELEMENTARY PRINCIPLES

OF

SCIENTIFIC AGRICULTURE..

BY

N. T. LUPTON, LL. D.,

PROFESSOR OF CHEMISTRY IN VANDERBILT UNIVERSITY, NASHVILLE
TENNESSEE.

NEW YORK ·:· CINCINNATI ·:· CHICAGO
AMERICAN BOOK COMPANY.

PRIMER SERIES.

SCIENCE PRIMERS.

HUXLEY'S INTRODUCTORY VOLUME.
ROSCOE'S CHEMISTRY.
STEWART'S PHYSICS.
GEIKIE'S GEOLOGY.
LOCKYER'S ASTRONOMY.
HOOKER'S BOTANY.
FOSTER AND TRACY'S PHYSIOLOGY AND
    HYGIENE.
GEIKIE'S PHYSICAL GEOGRAPHY.
LUPTON'S SCIENTIFIC AGRICULTURE.
JEVONS'S LOGIC.
SPENCER'S INVENTIONAL GEOMETRY.
JEVONS'S POLITICAL ECONOMY.
TAYLOR'S PIANOFORTE PLAYING.
PATTON'S NATURAL RESOURCES OF THE
    UNITED STATES.

HISTORY PRIMERS.

WENDEL'S HISTORY OF EGYPT.
FREEMAN'S HISTORY OF EUROPE.
FYFFE'S HISTORY OF GREECE.
CREIGHTON'S HISTORY OF ROME.
MAHAFFY'S OLD GREEK LIFE.
WILKINS'S ROMAN ANTIQUITIES.
TIGHE'S ROMAN CONSTITUTION.
ADAMS'S MEDIAEVAL CIVILIZATION.
YONGE'S HISTORY OF FRANCE.
GROVE'S GEOGRAPHY.

LITERATURE PRIMERS.

BROOKE'S ENGLISH LITERATURE.
WATKINS'S AMERICAN LITERATURE.
DOWDEN'S SHAKSPERE.
ALDEN'S STUDIES IN BRYANT.
MORRIS'S ENGLISH GRAMMAR.
MORRIS AND BOWEN'S ENGLISH
    GRAMMAR EXERCISES.
NICHOL'S ENGLISH COMPOSITION.
PEILE'S PHILOLOGY.
JEBB'S GREEK LITERATURE.
GLADSTONE'S HOMER.
TOZER'S CLASSICAL GEOGRAPHY.

LUPTON—SCI. AGR.

COPYRIGHT, 1880, BY D. APPLETON & CO.

W. P. 5

PREFACE.

THE following communication explains the reason that induced the author to present to the public this little work on "The Elementary Principles of Scientific Agriculture":

NASHVILLE, TENN., *December* 17, 1879.

In accordance with Chapter CLXXXVI., Acts of the General Assembly of Tennessee, approved March 27, 1879, which directs "that the Superintendent of Public Instruction and the Commissioner of Agriculture shall be constituted a commission to procure the preparation of, or the designation of a work on the Elementary Principles of Agriculture which shall be taught in the public schools of the State, as are the other studies prescribed in the 21st section of the Public School Law," the undersigned have "procured the preparation" of a work as herein described, by Professor N. T. Lupton, Professor of Chemistry, Vanderbilt University; and, having carefully examined the MS., hereby approve the same and adopt it, to be taught in the public schools of the State, in accordance with the express terms of said act.

(Signed)

Commis- { LEON TROUSDALE, *State Superintendent,*
sioners. { J. B. KILLEBREW, *Com. of Agriculture.*

In response to this demand for the introduction of the " Elementary Principles of Agriculture " into the regular course of study in the public schools, the author has endeavored to present the subject in clear, concise language, avoiding technical terms, except where scientific accuracy required their use. As the principles discussed are of universal application, this little work is designed for general use, being adapted to every section where agriculture is taught and practiced as a science. It is believed that intelligent farmers and planters will find it sufficiently plain and practical for profitable reading and study.

The attention of teachers is called to a few simple experiments in the Appendix, to which reference is made in the text by corresponding figures inclosed in parentheses. The performance of even a few experiments will add greatly to the interest of pupils in the facts and principles presented.

CONTENTS.

ELEMENTARY PRINCIPLES

OF

SCIENTIFIC AGRICULTURE.

CHAPTER I.

THE DEVELOPMENT OF SCIENTIFIC AGRICULTURE.

1. AGRICULTURE is both a science and an art. As an art, it teaches how to cultivate the soil, make and use fertilizers, take care of stock, and do whatever is necessary for the successful management of a farm. As a science, it explains the growth of plants and animals, and the principles upon which the practical operations of farming depend. As an art, it tells what to do ; as a science, it explains the reasons for what is done.

2. Agriculture, like other sciences, was practised as an art long before the principles upon which it is based were understood. As an industrial pursuit, it has always been the first in importance. This is owing to the fact that man's necessities compel him to cultivate the soil. While the earth of its own accord

produces enough food for the support of the lower animals, man is forced to earn his bread by the sweat of his brow.

3. In the early ages of the world, the population was small, and the wants of men so few that very little cultivation of the soil was required. The rivers and forests supplied, in a great measure, both food and clothing, until the increase of population originated necessities and luxuries which could only be satisfied by increased cultivation.

4. Although agriculture, as an art, has been practised to some extent by all nations, and in every age of the world, its progress as a science has been very slow. In fact, sciences of recent origin have made much greater progress and are at this day more generally understood. There are men even now who say that **practical** and **scientific** farming are very different; in other words, that agriculture is no science at all. This is because they do not understand its principles, or have a false notion of what scientific agriculture really is.

5. There are several reasons for the slow progress or growth of agriculture as a science. In the first place, the dignity and importance of the pursuit were not fully recognized until within recent times. It is true that, in every age of the world, some good and great men have been farmers. We read in history of Cincinnatus leaving his plow at the call of his country, and of Putnam deserting his field of agricultural labor for one of military glory, and even of

Washington retiring to Mount Vernon to spend the closing years of a noble life in the quiet occupation of a farmer. We read, too, of Plato, and Pliny, and Columella, and even of Cicero, living at times on their farms and writing on subjects connected with farming operations. These noted men, however, did not practise farming as a **profession**, as a **permanent** business, but rather as a relaxation from the cares of political life. To accomplish great results in any department of science requires that we make it more than a **recreation**, more than a **pastime**. It must be the study of a lifetime.

6. Again, agriculture has made slow progress as a science, because it is closely connected with and dependent on other sciences which are themselves of recent origin. A good farmer must know something of Botany, the science of plants, whether he finds it out by observation or learns it from books, that he may understand the character of the various products of the soil, and be enabled to adapt his mode of cultivation to the nature of his crop.

7. The farmer must know, at least practically, something of Zoölogy, the science of animals, that he may procure and raise such stock as will be of most service to himself, and command good prices in the market. He must be something of a geologist, that is, must know enough of Geology, the science that treats of the structure and formation of the earth, its soil and rocks, to enable him to understand the nature of soils, and be able to judge of their value by

mère inspection.   If not a mechanic himself, the mechanic arts must be understood by those who furnish him with plows and wagons and other farming implements.   The science of Physics, which treats of the general properties of bodies, and the causes—such as light, heat, and electricity—that modify these properties, is also important; and, finally, the farmer ought to have some knowledge of Chemistry, in order to understand the constitution of soils, and plants, and fertilizers, and to enable him to adapt them to each other, and, when necessary, to determine what ought to be added to a soil to make it produce a good crop.

8. We do not mean to say that the farmer or planter must be skilled in all these sciences, but he should have at least a general acquaintance with them if he would practise farming to the best advantage.   To the farmer, they are all profitable subjects for study, because they are closely connected with his profession, and are, in fact, the foundation upon which the science of agriculture is based.

9. Perhaps the most important practical results to agriculture have been derived from the science of Mechanics and that of Chemistry.   The progress in the mechanic arts is seen when we compare the farming implements now in use with those of former times.   The rough, uncouth wooden plow has given place to various forms of neat, easy-running plows partly or wholly of iron, and even these have given place in some countries to the gang and steam

plow, which do the work of a dozen old-fashioned implements. The sickle has given place to the cradle, the cradle to the reaper, the scythe to the mower, the old tedious process of separating cotton-seed from the fiber by hand, to the cotton-gin, and a great many other mechanical improvements which the inventive genius of the nineteenth century has given to the world.

10. Among the benefits derived from chemistry may be mentioned :

1. It teaches the composition and qualities of soils, of plants, of the atmosphere, of fertilizers.

2. It determines the kind and quality of food that different plants need for strong, healthy growth.

3. It shows how to manufacture fertilizers, and to make use of all sorts of refuse matter in the preparation of food for plants.

4. It explains the action of light, heat, and other agencies in promoting growth, and, in a word, unfolds all the conditions of fertility.

11. The chemist in his investigations proceeds to analyze the soil, and plant, and air, and fertilizers, that he may understand their nature and the relations they sustain to each other ; and we propose to inquire into the results of these investigations, and show in few words what modern science teaches of the composition and use of these substances with which the farmer and planter have so much to do.

CHAPTER II.

THE ORIGIN, COMPOSITION, AND CLASSIFICATION OF SOILS.

12. THE earthy matter in which plants grow is called **the soil.** It consists of finely divided particles of rock mixed with organic matter. The organic matter comes chiefly from the decay of plants, and forms generally a very small portion of the soil. It can be easily burned off, leaving the mineral portion, which consists of disintegrated or finely crumbled rock (1). The earth is supposed by geologists to have once been a mass of melted matter. As it cooled down, a crust was formed upon which condensed moisture falling as rain wore away the solid rock, being aided in its corroding action by heat and other agencies, until the pulverized or finely divided surface was in proper condition to produce the plant.

Soils consist of a small amount of organic matter mixed with rock which has been reduced to a fine state of division by mechanical and chemical agencies.

13. **Mechanical agencies** are such as merely alter the form or appearance of bodies without changing their character, like the grinding or rubbing of rocks together, whereby they are broken into small pieces. Each little piece has the same properties as the large mass. **Chemical agencies** are

such as alter the real nature of bodies, as in the rusting of iron and the burning of wood. The rust has different properties from the metallic iron, and the gases and ash formed in burning are very different from the original wood. In such cases, chemical action has brought about the change. Both these agencies are all the time at work upon the rocks, and, though they are generally slow in their action, vast results are finally produced.

14. The geologist includes under the term **rock** all soils and loose material, clays, and gravel, as well as the large solid masses which compose the earth. The disintegrating, or crumbling action by which soils are formed is always going on, plants and animals doing a great deal toward bringing about the changes that are produced. The most powerful agencies in the crumbling of rocks are air and water. Water not only wears away rocks and reduces them to powder by mechanical action, but it dissolves them all in a greater or less degree. This solvent action is greatly increased by the air which it contains. Water also soaks into soft rocks and runs into holes and crevices, where freezing it expands, and forces masses and particles of rock asunder. One great advantage of deep, thorough cultivation is that air and moisture may penetrate the soil, and continue their disintegrating or crumbling action upon the particles of rock of which the soil is composed, thus setting free the plant-food and preparing it as proper nourishment for the hungry plant.

15. The elements of a soil vary with the kind of solid rock from which it was originally formed. A disintegrated sandstone, or limestone, or slate, or granite, will each produce a soil of peculiar character. Very often the soil rests upon the rock from which it was formed. Sometimes, however, it has been carried away and deposited at a long distance from the parent rock. Such deposits form what are called **alluvial** soils, and are generally found in creek and river bottoms (2).

16. The great difference in the quality and value of soils results mainly from a difference in the relative quantity of certain elements present. The total absence of any of the essential elements of plants is rare. Notwithstanding the great variety of mineral and vegetable substances, the whole mass of the earth and everything upon it is composed of few **elements,** or **simple** substances (3).

17. The number of elements certainly known is **sixty-five**; several others have been announced recently, but their existence is yet in doubt. Five of these, oxygen, hydrogen, nitrogen, chlorine, and fluorine, are gases, and two, bromine and mercury, are liquids at the ordinary temperature of the air, while all the others are solid. By reducing the temperature and applying great pressure, the gases can be changed to liquids, and the liquids can be converted into solids. There are no **permanent** gases; even oxygen, hydrogen, and nitrogen have been made to assume the liquid form.

18. The farmer and planter are concerned with only fourteen or fifteen of these elements, because only this number enter into the composition of soils generally, and are concerned in the growth of plants. These are as follows:

Non-metallic Elements.	*Metallic Elements.*
1. Oxygen.	1. Potassium.
2. Hydrogen.	2. Sodium.
3. Nitrogen.	3. Calcium.
4. Carbon.	4. Magnesium.
5. Silicon.	5. Aluminium.
6. Sulphur.	6. Iron.
7. Phosphorus.	
8. Chlorine.	

To these may be added manganese, iodine, and fluorine, which are sometimes present in minute quantities.

19. **Oxygen** is by far the most abundant of the elements. It forms about one half of the solid crust of the earth, eight-ninths of all the water, and one-fifth of the atmosphere. This element is easily prepared by heating in a glass tube or flask either mercuric oxide (commonly called red oxide of mercury), or potassium chlorate. Heat separates the oxygen from these compounds. It can easily be collected in suitable vessels and its properties examined. If a small quantity of manganese dioxide (black oxide of manganese) be mixed with the potassium chlorate, the oxygen will be given off at a lower temperature. (4).

20. Oxygen is an invisible gas without taste or smell. It is called a supporter of combustion because wood, coal, oil, gas, and other substances burn in it. If the oxygen be pure they burn with great brilliancy. Combustion, such as takes place in our fireplaces, stoves, and lamps, is the result of chemical union between the oxygen of the air and the carbon and hydrogen of the fuel. The bright sparks that fly from the blacksmith's anvil are particles of iron uniting with oxygen.

21. Oxygen unites to form compounds with all the known elements except fluorine. These compounds are called **oxides.** The compounds of oxygen with some of the metals were originally named by merely changing the termination **um** or **ium** of the metal into **a.** These names are still in common use. Thus:

Potassium hydroxide is often called potassa or potash.
Sodium hydroxide   "   "   soda.
Magnesium oxide   "   "   magnesia.
Aluminium oxide   "   "   alumina.

Calcium with oxygen forms lime; silicon forms silica or pure white sand. According to the new system of naming oxides, the name of the metal comes first with the word oxide immediately after.

22. Oxygen sometimes unites slowly and gradually with other elements without producing light or intense heat, as when wood decays or iron rusts. These are cases of **oxidation,** and the final results

are the same whether the action takes place slowly or
rapidly. This process of slow oxidation is constantly
going on in the bodies of animals. When they breathe,
the air enters the lungs, where the oxygen it contains
is taken up or absorbed by the blood, and carried
throughout the body. Animals cannot live without
oxygen. It is also necessary to the growth of plants.

23. **Hydrogen** is another abundant element. It
forms one ninth by weight of water and enters into
the composition of all plants and animals. It can
be prepared easily by the action of dilute sulphuric
or hydrochloric acid on zinc or iron. If some scraps
of either of these metals be placed in a wineglass
and a little acid poured over them, effervescence will
be produced by the escape of the hydrogen (5).

24. Hydrogen is the lightest substance known.
Like oxygen it is an invisible gas without color, taste,
or smell, but unlike oxygen it will burn when brought
in contact with flame. The flame of burning hydrogen is of a pale-blue color and intensely hot (6).
The result of the combustion is water (7). Sixteen
pounds of oxygen unite with two pounds of hydrogen to form eighteen pounds of water.

25. **Water,** which is formed by the chemical
union of the two elements just mentioned, is one of
the most abundant and important compounds in
nature. It is easily converted into vapor, which, ris-
ing from the surface of the earth in an invisible form,
is afterward condensed into rain, and dew, and
frost. It dissolves solid, liquid, and gaseous sub-

2

stances, and in this way carries food to the roots of plants (8). All natural waters, such as are found in springs, wells, rivers, and lakes, contain in solution more or less mineral matter derived from the soil. Rain-water is the purest, as it contains nothing in solution except what it gets from the air as it falls through it.

26. **Nitrogen** is found in the atmosphere in large quantity, and also in plants and animals. It forms four-fifths of the volume of the air. The most convenient way to prepare nitrogen is to remove the oxygen from air by means of phosphorus (9). Other substances can also be used for the removal of the oxygen.

27. Nitrogen is a colorless gas with no taste or smell. It neither burns nor does it support combustion. Animals cannot live in it, and yet it is not poisonous. It has no power to sustain life, merely serving in the atmosphere to dilute the oxygen. The three gases just mentioned are easily distinguished by means of a lighted taper. Oxygen will not burn, but will make the taper burn with great brilliancy; hydrogen will put out the taper, but will burn itself with a pale-blue flame; nitrogen will neither burn nor will it support the flame, but put it out at once.

28. **Ammonia** is a very important compound of nitrogen and hydrogen. It is formed when ammonium chloride (sometimes called sal-ammoniac) is rubbed with common lime, and is a gas of strong, pungent odor (10). It unites with acids, destroying

their sour taste. Water absorbs about seven hundred times its volume of this gas, and in this condition it is commonly called **hartshorn.** Ammonia is an **alkali**—that is, it neutralizes acids and restores vegetable colors, like litmus, that have been made red by means of an acid.

29. Ammonia is formed naturally by the decay of animal and vegetable matter, as in manure-piles. Its odor can be detected in stables, or where organic substances containing nitrogen are undergoing decomposition. As a gas it exists in minute quantities in the atmosphere, and finds its way into plants through their roots.

30. Nitrogen and oxygen unite to form several oxides, one of which when combined with water is known as nitric acid. This is a very strong acid of intensely sour taste. It corrodes or destroys the flesh, and acts upon nearly all the metals, forming with them a class of compounds, called **nitrates.** One of these, potassium nitrate, is known as nitre or saltpetre; and another, sodium nitrate, goes by the name of Chili saltpetre. Both of these nitrates are used as fertilizers (11).

31. **Carbon** is an element that exists in three distinct forms. Charcoal, coke, and lampblack are varieties of the first; graphite, or plumbago, commonly known as **black lead** and used for making lead-pencils, is the second; and the diamond, which is pure crystallized carbon, the third and most valuable form. When wood is heated in a close ves-

sel or burned in a covered heap, the black residue is
chiefly carbon. This element forms the greater portion of woody substances and enters largely into the
composition of all organic matter. Sugar contains
42 per cent. of carbon, spirits of turpentine 88 per
cent. The black smoke of a candle or lamp is carbon in a finely divided state (12).

32. **Carbon dioxide**, or **carbonic acid gas**,
is a compound of carbon and oxygen that exists in
small quantities in the air, and is always formed
when carbon burns. It is a heavy, poisonous gas
that is formed during respiration, combustion, fermentation, and decay. While it poisons animals, it
is an important food for plants, as will be shown
hereafter. If a little hydrochloric acid be poured
on a piece of marble or limestone, the effervescence
produced is caused by the escape of this gas (13).
It will extinguish a burning taper.

33. **Silicon** is an element found in common sand
and quartz. It is very abundant, forming about one-
fourth of the solid part of the earth, but very difficult to separate from the oxygen with which it is nearly always combined. Some chemists doubt whether
it is a necessary constituent of plants. It is certainly
found in the stems of grass, wheat, corn, and in various kinds of vegetation.

34. **Sulphur** is a well-known substance of yellow
color that burns with a pale-blue flame and suffocat-
ing odor. In burning, it unites with the oxygen of
the air to form sulphur dioxide, which is used to de-

stroy bad odors and also for bleaching straw bonnets, hats, and other goods (14). Sulphur in combination with oxygen and hydrogen forms sulphuric acid, or "oil of vitriol," one of the strongest of the acids. This acid forms compounds known as sulphates, as calcium sulphate, or gypsum, and magnesium sulphate, or Epsom salts (15).

35. **Phosphorus** is a soft, slightly yellow solid, that gives off a white smoke when exposed to the air and takes fire very easily. This white smoke is caused by the slow burning of the phosphorus (16). This element forms a large percentage of the bones of animals, and is found in soils and in plants, especially in the seeds. Phosphorus takes fire so easily in the air that it must be kept under water. It is used for making matches. When the end of a match is rubbed against a rough surface, enough heat is produced to set fire to the phosphorus; this inflames the sulphur on the match, and the sulphur sets fire to the wood of which the match is made (17).

36. **Phosphoric acid** is a compound of phosphorus with oxygen and hydrogen (18). When calcium takes the place of hydrogen in this acid it forms calcium phosphate, or bone phosphate, the chief constituent of bones. This acid forms a large class of salts called **phosphates**. As animals get their food from plants and plants their food from the air and soil, it is necessary for soils to contain phosphorus. When a deficiency of this element does exist in a soil, it is usual to supply it by means of bones.

37. **Chlorine** is a heavy gas of a yellowish-green color, found only in combination with other elements. In combination with hydrogen it forms hydrochloric acid, formerly called muriatic acid. If this acid be heated with manganese dioxide, the chlorine will be separated. It is poisonous when breathed. It will destroy organic coloring matters and bad-smelling gases, and hence is used for bleaching, and as a disinfectant or purifier of air (19). Common salt is a compound of sodium and chlorine. This element is found in the ash of plants and also in soils.

38. **Iodine**, a dark-colored solid that forms a beautiful violet vapor when heated, and **bromine**, a dark-red liquid, are found in combination with other elements in some plants. In chemical properties they are similar to chlorine (20). **Fluorine** is the most difficult to prepare of all the elements. It is described as a gas, and exists in small quantities in the teeth of animals, and in other portions of the body, and also in some plants and minerals.

39. **Potassium** is a soft metal, lighter than water. It unites with oxygen so readily that it has to be kept under the surface of naphtha, a liquid that has no oxygen in it. If thrown on the surface of water, or placed on a piece of ice, it takes fire and burns with a beautiful violet-colored flame (21). In combination with hydrogen and oxygen, it forms caustic potash. All the acids contain hydrogen. The salts of potassium are formed by replacing the hydrogen of

an acid with this metal. Some of these salts are found in soils and in the ashes of plants.

40. **Sodium** is a soft metal very much like potassium in appearance and in properties. It also must be kept under naphtha, but will not take fire on water, unless it be held in one place, or the water be warmed. It spins around over the water, decomposing it, and forming caustic soda (22). It forms a large number of salts. Common salt is sodium chloride, and is found in all salt springs, in the ocean, and in soils, and the ashes of plants. Caustic soda and, caustic potassa, or potash, are called **alkalies.** They destroy the flesh, neutralize acids, and color red litmus-paper blue (23). In these respects they are similar to ammonia. They are used to make soap : caustic potash to make soft soap, and caustic soda to make hard soap.

41. **Calcium** is a metal very hard to separate from its compounds. With oxygen it forms common, unslaked lime. Marble, and limestone, and chalk, are calcium carbonates, which on being heated lose carbon dioxide, and are changed into lime. Gypsum, or land plaster, is calcium sulphate, a valuable fertilizer. Spring and well water very often contain these salts in solution.

42. **Magnesium** and **aluminium** are hard, white metals, the former a constituent of dolomite, or magnesian limestone, and some other rocks; the latter is found in all clay and slate rocks. The metal magnesium burns with great brilliancy and is sometimes

used for lighting up caves (24). Aluminium has been used to some extent for ornamental and other purposes, but it is so difficult and expensive to separate from its compounds that it has not come into general use.

43. **Iron** is a common metal with which everybody is familiar. Its ores, such as hematite and limonite, are abundant, and are used in immense quantities for the manufacture of this useful metal. It is found in all soils forming the coloring matter of clays, and exists in a great variety of minerals. Manganese is a metal very much like iron in its chemical properties, but much more difficult to separate from its ores.*

44. In giving the composition of a soil it is usual to state the percentage weight of the metallic oxides and acid-forming oxides, as it is known that the elements do not exist in the soil in a free state. The process of analyzing a soil, that is, of finding out its constituents, is easily understood **in theory**, but to perform the analysis requires some practice and skill in chemical work. Such analyses are not so useful to the farmer as was once supposed, since the **condition** of the elements in the soil has more to do with its fertility than **quantity.**

45. The following analysis of a fertile soil will give some idea of the relative quantity of the constituents usually found:

---

* For a more extended description of these elements the student is referred to works on chemistry.

Per cent.

Potassium oxide	0.2
Sodium oxide........................	0.4
Calcium oxide, or lime....................	5.9
Magnesium oxide, or magnesia...............	0.8½
Iron oxide, or ferric oxide..................	6.1
Aluminium oxide, or alumina...............	5.7
Manganese oxide........................	0.1
Silicon oxide (silica).....................	64.8
Sulphuric acid, or sulphur trioxide...........	0.2
Phosphoric acid, or phosphorus pentoxide......	0.4½
Carbonic acid, or carbon dioxide.............	4.0
Chlorine........................	0.2
Organic matter.........................	9.7
Loss............	1.4
Total..............................	100.0

The proportion of some important constituents, as potassium oxide, phosphoric acid, etc., is very small, but the smallest would amount to several tons per acre.

46. Silica, or sand, is the chief substance in a great many soils. Clay, a compound of silica and alumina with small quantities of other elements, is also abundant in most soils; so are lime and iron oxide. The amount of organic matter is quite variable. In some soils it reaches 15 or 20 per cent., while in others it is less than 1 per cent. The total absence of any of the constituents mentioned above, except perhaps aluminium oxide and manganese, would render the soil unproductive for ordinary crops. Those in least abundance, as potassium oxide and phosphoric acid,

are most likely to be deficient, and in such cases they must be supplied by the use of fertilizers or manures.

47. The great quantity of sand in most soils and its presence in all have suggested the propriety of classifying soils according to the amount of sand they contain, as follows:

1. **Pure clay,** from which no sand can be removed by washing.

2. **Strong clay,** when the soil contains from 5 to 20 per cent. of sand.

3. **Clay loam,** when it contains from 20 to 40 per cent. of sand.

4. **Loam,** from 40 to 70 per cent. of sand.

5. **Sandy loam,** from 70 to 90 per cent. of sand.

6. **Light sand,** more than 90 per cent. of sand.

It is easy to classify soils in this way by merely washing out the sand and weighing it (25).

48. When soils contain a large amount of calcium carbonate, they are said to be **calcareous,** or **marly;** and, when a very large percentage of organic matter, they are said to be **peaty,** or are called **vegetable mold.** The presence of a large quantity of clay makes a soil sticky when wet, and causes it to hold moisture a long time, hence such soils are said to be **heavy;** a large quantity of sand gives the opposite property, that is, of not retaining moisture, and hence these are said to be **light.**

49. The **soil** proper is the surface layer down to

where a change in the character of the material takes place, generally from six to ten inches and beneath this is the **sub-soil.** Deep cultivation increases the depth of the soil, and allows air and moisture to enter, and the roots of plants to penetrate farther in · search of food.

CHAPTER III.

THE COMPOSITION OF PLANTS.

50. THE following **ten** elements are always found in plants, and are believed to be absolutely essential to their growth :

Non-metallic.	*Metallic.*
1. Carbon.	7. Potassium.
2. Hydrogen.	8. Calcium.
3. Oxygen.	9. Magnesium.
4. Nitrogen.	10. Iron.
5. Sulphur.	
6. Phosphorus.	

Four others, sodium, manganese, silicon, and chlorine, are generally found, and in marine, or sea plants, iodine and bromine. Traces of fluorine have also been found, and sometimes minute quantities of lithium, cæsium, and rubidium, and a few other elements.

51. A great many analyses and experiments have been made by chemists to find out just what ele-

ments enter into the structure of plants, and whether they are all really necessary for plant-growth. The results of these investigations, as given in the last section, show what sort of food plants must have in order to grow. Aluminium, an element found in all soils and the base of all clay, does not enter into plants. As animals derive their food from plants or from each other, the same elements, except **silicon,** enter into the composition of both.

52. If a plant be heated up to the boiling-point of water, it loses a large part of its weight by the escape of water which it contains, and becomes dry. In turnips and cabbages, nine-tenths, and in potatoes three-fourths of their weight is water which can be driven off by heat. Even in cured hay and fodder, from one-sixth to one-tenth is water. If the dried plant be exposed to a red heat, it will take fire and the greater part be consumed. The small portion left is called the **ash,** and is generally a fine white or gray powder (26).

53. The portion of a plant that burns is chiefly carbon combined with some hydrogen and oxygen, and a little nitrogen. These four elements are sometimes called **organic** elements, because they form by far the larger part of organic bodies. In the process of combustion, while these elements disappear, it must not be supposed that they are destroyed. They merely enter into new combinations with the oxygen of the air, forming chiefly carbon dioxide and watery vapor, which float off unseen in the atmosphere.

54. When plants are burned with free access of air, the sulphur and phosphorus are oxidized, and remain in the ash with the metals. Chlorine and silicon are also found in the ash. This ash, or inorganic matter, varies in quantity from about one-half to one per cent. in some kinds of wood to from 15 to 18 per cent. in tobacco. The following table shows about the average percentage of ash in a number of vegetable products :

Percentage of Ash in Dried Products.

Cotton, lint	1.0	Red clover	6.8
" seed	8.9	Cabbage	8.0
Wheat, grain	1.9	Irish potatoes	4.3
" straw	5.0	Turnips	10.0
Indian corn	1.5	Tobacco	15–18

The elements that enter into the composition of the ash are given in the following table :

Percentage Composition of the Ash of Plants.

	WHEAT.		Indian corn.	Potatoes.	Red clover hay.	Tobacco.*
	Grain.	Straw.				
Potassium oxide	31.54	12.16	37.95	61.60	31.86	32.63
Sodium oxide	2.66	1.00	3.00	1.00	2.16	3.81
Magnesium oxide, or magnesia	12.10	4.00	7.50	5.00	12.16	12 10
Calcium oxide, or lime	3.14	6.82	3.40	2.40	31.03	40.15
Iron oxide	trace.	1.02	0.40	0.85	0.66	.....
Phosphorus pentoxide, or phos-phoric acid	48.50	3.23	44.80	17.67	9.00	3.74
Sulphur trioxide, or sulphuric acid.	0.08	5.78	1.50	6 25	3.03	4.02
Silicon dioxide, or silica	1.88	65.34	1.45	1.00	6.71	2.69
Chlorine	0.10	0 60	trace.	2.23	3.33	0.86
	100.00	100.00	100.00	100.00	100.00	100.00

* Average of several analyses of Kentucky tobacco made in the laboratory of Vanderbilt University by Messrs. Hobbs and Wooldridge.

55. The elements heretofore mentioned as forming the material of plants are sometimes called **ultimate** elements, because they are the simplest forms of matter into which plants can be separated, or resolved. What are called the **proximate** elements, or principles, are compound substances, such as starch, sugar, gum, oil, cellulose, or woody fibre, etc. While there are a large number of these, each species of plant having its peculiar principle, the greater number can be included under the following groups.

56. **Amylaceous** and **saccharine** substances, such as starch, sugar, cellulose, or woody fibre, and gum. These are composed of only three elements, carbon, hydrogen, and oxygen. The woody fibre, or cellulose, forms the greater part of many plants, and usually consists of small tubes sticking to each other. Cotton fibre is nearly pure cellulose. It is contained in all plants, in the stems, leaves, roots, and seeds. It has the same elements in the same proportion as starch, and differs in these respects very slightly from sugar. It can be converted into one kind of sugar, **glucose**, by boiling with dilute sulphuric acid.

57. **Pectose** substances, like the jellies and pulp of fruits, and of some roots, as the turnip, beet, and onion. These substances are closely related to woody fibre, and are easily changed into it by the plant.

58. **Vegetable acids,** such as tartaric acid found

in grapes, citric acid in lemons, and malic acid in apples. These vegetable acids are numerous, but all have the same elements, carbon, hydrogen, and oxygen, that are found in sugar.

59. **Fats** and **oils**, such as are found in olives, cotton-seed, flax-seed, etc. These are composed of the same elements as sugar, but have proportionally less oxygen. Turpentine, resin, and different kinds of wax, are included in this group.

60. **Albuminoid**, or **protein** bodies, which differ from the group mentioned, in having nitrogen as a constituent, and in some cases sulphur and phosphorus. Albumen is found nearly pure in the white of an egg, and a similar substance is contained in the juices of plants. The term "albumen" is used by botanists to mean any nutritive material found in the seeds of plants without reference to its chemical composition. Chemists apply the word only to substances containing nitrogen, and in this sense it is here used. Gluten, a sticky substance found in flour, belongs to this class, and vegetable casein, similar to the white curd of milk, found in leguminous plants, such as the pea and bean.

61. Notwithstanding the great variety of these **proximate** principles, and their great difference in physical and chemical properties, they are composed of only a few elements. A knowledge of their chemical constitution has very greatly simplified the changes which take place in the growth and maturity of plants, and shown that a very slight alteration in chemical

constitution will convert a disagreeable or tasteless fruit into one that is sweet and delicious.

62. Starch and sugar differ from each other very slightly in the number and proportion of their constituent elements, while starch, cellulose, and gum are the same in these respects. How the internal structure or arrangement of these elements causes a change in properties, we do not know. Starch forms the larger portion of the seeds of most plants, and especially of those used as food. Wheat, corn, rice, and potatoes, when dry, are about two-thirds starch.

63. Grape-sugar differs from starch by one equivalent of water, that is, by the addition of a little water chemically combined, starch becomes sugar. This change takes place in seeds to some extent in the process of sprouting, and is taken advantage of in the manufacture of beer and whisky from barley and corn. A lock of cotton, or even some sawdust, can be changed into sugar suitable for manufacturing alcohol by dissolving it in strong sulphuric acid, diluting with water, and boiling for some time.

64. Every green leaf is a tiny laboratory, where by the aid of the sun's rays those changes are made which give us wood, and starch, and sugar, and gluten, and gum. To make these the plant uses carbon dioxide and ammonia, and water and other substances, which enter the plant as food.

65. The following table gives the composition of various crops:

	Water.	Albu-min-oids.	Fat.	Non-ni-trogen-ized ex-tractive matter.	Woody fibre.	Starch.	Ash.
Grasses (hay)................	15.0	9.4	2.6	38.8	28.5	....	5.7
Red-clover hay	15.0	14.2	3.1	37.2	24.8	....	5.7
Wheat......................	15.0	12.0	1.5	....	5.0	67.0	1.7
Oats.......................	14.0	11.5	6.0	....	9.0	56.5	3.0
Indian corn	14.5	10.0	7.0	....	5.2	61.4	1.9
Buckwheat.................	14.0	9.0	2.5	....	12.0	60.2	2.3
Rye......	16.0	9.0	2.0	....	8.0	64.0	1.0
Rice......,,....	14.0	5.3	1.0	....	2.5	76.7	0.7
Cotton-seed...............	6.6	31.9	31.2	...	7.3	14.0	9.0
Potatoes..................	75.5	2.1	0.7	....	2.1	18.6	1.0

66. The ash constituents, though small and vari-able in quantity, are of prime importance in the growth of the plant; and just here we find the whole theory of fertilizing. The moisture and volatile com-pounds formed during combustion float off unseen in the air when a plant is burned or decays, and, by this same air where these compounds always exist, they are carried to other plants and serve as proper food, while the ash constituents can not be thus trans-ported. Every plant removed from the soil carries with it the elements of this ash, and, as some of these elements exist in minute quantities, the soil is sure to become exhausted where means are not used to have them restored.

67. The old system of carrying away and never restoring these necessary elements has well been called by Liebig **a system of spoliation.** Some elements, such as iron and silicon, and, in many soils, calcium and sodium, are so abundant that, practically, they may never become deficient, and yet they may

3

not be in a condition to be readily used as food by the plant. To remedy this condition, either **mechanical** means must be used, such as repeated plowing, deep and thorough, to bring about exposure to air, and rain, and sunshine, and frost ; or **chemical** means, as the application of lime and various salts which supply food directly, or which exert a decomposing action, and thus prepare the food which already exists in the soil and render it suitable for use.

68. Phosphorus and potassium, which are generally in minute quantities, as well as other elements when deficient, must be supplied, or the vigorous growth of plants whose structure requires these elements—and this includes all of our valuable crops—is an **utter impossibility.** As well might the carpenter dispense with nails in the construction of a house, or the tailor with thread in making a coat, as the farmer dispense with these essential elements in the growth of his crop.

69. As soils are of great variety, their composition, or character, should be determined, and the deficiency pointed out, if any exists, before an attempt be made to remedy the defect. While chemistry teaches the farmer the cause of the exhaustion of his soil and shows how this can be prevented, it must not be held responsible for the failures resulting from the use of the many mixtures which are sold under various names as fertilizers.

70. The means used for furnishing plants with the elements necessary for their growth, that is, for

keeping a soil fertile, whether mechanical or chemical, must be guided by sound judgment. While practical observation and experience teach us that lands do wear out by continued cultivation and removal of crops, chemistry explains the cause, and points out the remedy. This may be by the purchase of concentrated fertilizers, or by the use of such fertilizing materials as already exist or can be produced on every farm. This subject will be discussed in a subsequent chapter.

CHAPTER IV.

COMPOSITION AND PROPERTIES OF THE ATMOSPHERE.

71. THE atmosphere from which plants obtain a large portion of their food is a mixture of oxygen and nitrogen in the proportion of about one-fifth of the former to four-fifths of the latter, with a small quantity of carbon dioxide, a trace of ammonia, a variable quantity of watery vapor, and traces of a few other gases resulting from combustion and decay Its composition by volume may be stated as follows

Nitrogen	77.95
Oxygen	20.61
Watery vapor	1.40
Carbon dioxide	0.04
Ammonia } Hydrogen carbide .. }	traces
In towns { Hydrogen sulphide. } { Sulphurous oxide... }	traces
Total	100.00

72. These constituents, though gaseous and invisible, can be separated and measured by the chemist as surely and definitely as the farmer can measure his corn and wheat, or the planter weigh his cotton and tobacco. Two of these, carbon dioxide and ammonia, though in very small quantities, are sufficient for the ordinary growth of plants. They are all so nicely intermingled, so uniformly mixed, that dry air, no matter where collected, is found to possess essentially the same composition.

73. Watery vapor, as above stated, is a variable constituent of the atmosphere. It rises continually from the surface of the land, as well as from every lake, and river, and ocean, and comes to us again in gentle showers which the cool currents of air condense, or in dew-drops which settle by night on leaf and flower. Beautifully harmonious are the laws and operations of nature, and especially that law by which the vapor of water so gently and continuously ascends from the earth's surface and, condensing, falls to be again vaporized after it has performed its part in the support of animal and vegetable life.

74. The study of chemistry has revealed the part that each of these constituents of the atmosphere performs in the economy of nature. **Nitrogen** is negative in its character, being indifferent toward entering into combination with other elements, while **oxygen** is active and energetic in supporting combustion, and in sustaining animal life. Our bodies, like stoves, are consumers of carbon and oxygen, and

produce in a similar manner the gas known as carbonic acid, or carbon dioxide. This gas is poisonous to animals when it exists in any considerable quantity in the air, but is absolutely essential to the growth of plants.

75. **Carbon dioxide** consists of carbon and oxygen, and is generated in all cases of ordinary combustion, in putrefaction, fermentation, and decay. It is also a product of respiration, and would soon accumulate in sufficient quantity to destroy animal life, were it not for the fact that it is absorbed by the leaves of trees and plants, there deprived of its carbon, and pure oxygen restored to the air whence it came (28).

76. Every one has observed the soft, porous texture of the under surface of the leaves of plants; through these pores the carbon dioxide is inhaled as it floats in the atmosphere, and under the influence of sunlight the process of digestion takes place within the plant whereby the carbon is retained as food, and the oxygen exhaled, thus preserving the purity of the air.

77. The relative amount of carbon dioxide in the atmosphere is only about one twenty-fifth of one per cent., and yet it is enough for the purposes of vegetation. A distinguished chemist says, "The gigantic trees which adorn the forests of tropical regions with the dense pine woods of more northern zones, and the abundant though less conspicuous vegetation of temperate climes, all derive their stock of carbon

from this small but essential constituent of the atmosphere."

78. Ammonia exists in the air in even more minute quantities than carbon dioxide, so minute that only the most delicate tests can indicate its presence, and the most sensitive balance would not be affected by the quantity in a room of ordinary size, and yet it is of essential value to all kinds of vegetation.

79. A close study of the nature and composition of the atmosphere shows that it is admirably adapted, by its physical and chemical properties, to the wants of animals and plants. "It conveys to them their nourishment and life; it tempers the heat of summer with its breezes; it binds down all fluids, and prevents their passing into a state of vapor; it supports the clouds, distills the dew, and waters the earth with showers; it multiplies the light of the sun, and diffuses it over earth and sky; it feeds our fires, turns our machines, wafts our ships, and conveys to the ear all the sentiments of language and all the melodies of music."

80. Science teaches us that Infinite Wisdom has shown, in these invisible atmospheric agencies at work around us, the same skill and benevolence, the same goodness and power, as in the more manifest displays of divine energy. Silently but surely they accomplish their appointed work of contributing directly to the nutrition of organic bodies, and by their action upon solid substances continually

prepare material to take its proper place in the building up of animal and vegetable structures.

81. The atmosphere acts upon both organic and inorganic matter so as to reduce it to simpler forms. As soon as an animal or plant dies, by contact with the air its elements quietly but surely undergo a change, recombining to form new and simpler compounds capable of entering again into organic bodies.

82. Knowing the composition of the soil, of plants, of air, we are prepared to study their relations and the laws of growth and development in plants.

CHAPTER V.

THE SOURCES OF PLANT-FOOD AND HOW OBTAINED.

83. THE earth and atmosphere are, of course, the only possible sources of food for plants, but the interesting questions arise, What portion of food does each furnish? and how is it taken up and used by the plant? If we include hydrogen, which is one of the elements in watery vapor and in ammonia, there are four elements in air, viz., oxygen, nitrogen, carbon, and hydrogen. A great many experiments have been made to find out whether these four elements enter directly from the air into plants, through the leaves, or are taken in by the roots from the soil.

84. **Carbon,** the most abundant element in organic substances, cannot enter the plant in a pure

state, as it is perfectly insoluble. In the gaseous form, as carbon dioxide, it is absorbed in large quantities by the leaves, and it also enters through the roots. By the action of sunlight this gas is decomposed in the green leaf, the carbon retained, and the oxygen restored to the air. This change cannot take place without the influence of sunlight, and hence plants grow much more rapidly in the daytime than at night. It is believed, by those who have studied the subject closely, that all the carbon contained in farm crops is derived from the atmosphere.

85. **Hydrogen** and **oxygen** enter in the form of water through the roots, and carry with them the various soluble matters of the soil that are needed for growth. Hydrogen also enters in the form of ammonia, which is a compound of nitrogen and hydrogen, and as hydrogen compounds formed by the decomposition of organic matters in the soil. In the same manner oxygen, in combination with carbon and with hydrogen and other elements, is absorbed in large quantities.

86. **Nitrogen** is also taken in as food by both leaves and roots, but always in combination. This element forms nearly four-fifths of the atmosphere, and yet, as free nitrogen, it does not contribute in the least to vegetable growth. The whole of the nitrogen is obtained from compounds of ammonia and from nitrates.

87. Plants contain very little nitrogen, generally from one half to three per cent. This, however, is

as necessary as those elements that enter largely into their composition. One ton of hay contains about thirty pounds of nitrogen, a small quantity proportionally, but in one hundred tons it amounts to a great deal. The stimulating effect of guano and similar manures is due mainly to the nitrogen they contain in the form of ammonia and its salts. While the atmosphere may contain ammonia enough to meet the ordinary demands of vegetation, it has not enough to supply the extraordinary demand of an immense crop in a limited time.

88. The four elements just mentioned are sometimes called **organic** elements, because they form much the larger part of all organic bodies. The remaining constituents, called **inorganic** elements, are found in the ash of plants, and are always taken in as soluble matter through the roots. The question arises, In what manner? Before answering this question, it will be well to examine the structure and use of the different parts of a plant.

89. The main parts of a plant are the **root,** the **stem**, and the **leaves** (29). The root spreads through the earth as the stem does through the air. The seed contains an **initial s**   that is, the beginning of a stem, with food enough   tart it to growing. It requires for germination or first growth, air, warmth, and moisture. The moisture which is first absorbed causes the seed to swell, oxyge enters, a chemical change takes place in the elements composing it whereby the germ, or little plant, gradually enlarges,

the covering bursts, and the radicle, or rootlet, appears, and buries itself in the earth, while the little stem or **plumule**, as it is called, rises to the surface of the ground in search of sunlight. Nature has stored up in the seed enough material to supply the growth thus far, but henceforth its food must come from the soil and atmosphere. This, in some mysterious way, through the influence of sunlight, it is enabled to take in and digest.

90. The root once started, divides and subdivides, sending out branches in every direction and hunting industriously in the soil for nourishment, while the stem and leaves do the same in the atmosphere. The leaves are in some sense the lungs of the plant, while the roots are its mouths. The length to which the roots sometimes extend is astonishing, and they really seem to have a kind of instinct that guides them to their proper food. Those which find proper nourishment enlarge and multiply rapidly, while those that do not, die or remain undeveloped. For a plant to be thrifty, it must have plenty of suitable food, and have it near at hand. It cannot, like an animal, run about in search of what it needs, though it will extend its roots to a long distance if necessary to get suitable material for growth.

91. Schubert, a German agriculturist, made an excavation in a field to the depth of six feet, and directed a stream of water against the vertical wall of soil until it was washed away, so that the roots of

plants growing in it were laid bare. Roots were exposed in this way in a field of rye, also in one of ·
beans, and in a bed of garden-peas, which were
found to present the appearance of a mat of white
fibres to a depth of four feet from the surface of the
ground. He found the roots of winter wheat as deep
as seven feet in a light sub-soil forty-seven days
after sowing. Another German, who has studied
the subject, calculated the total combined length
of the roots of a vigorous barley plant in a rich
garden soil to be one hundred and twenty-eight
feet, and in a coarse-grained, compact soil eighty
feet.

92. The absorbing surface of roots is greatly increased by very small root-hairs which are lost as
the roots become old. It was once supposed that
the ends, or tips of roots consist of delicate tissue,
or organs called " spongioles," through which alone
orption of food takes place, but such is not the
case ; no organ or structure of the kind exists.

93. Agricultural writers distinguish three kind of
roots, viz., **soil** roots, **water** roots, and **air** roots.
Nearly all our useful plants have soil roots which
perish if kept for a length of time in air or water,
while some plants, as rice for instance, have roots
which can grow either in soil or water. The common mulberry and China tree will extend a portion
of their roots down to the bottom of a well thirty or
more feet in depth, and mat over the bottom com-
pletely, while the cornstalk will put out air roots at

a joint above the ground, which extend until they · reach and penetrate the soil.

94. It was once supposed that roots have the power of **excretion**, the reverse of absorption, but Johnston says, " In the light of newer investigations touching the structure of roots and their adaptation to the medium which happens to invest them, we may well doubt whether agricultural plants in the healthy state excrete any solid or liquid matters from their roots." The food that enters a plant through the roots must be in a state of solution. There is no doubt that plants have, to a certain extent, the power of **selection**. We know that plants grown on the same soil and under the same conditions do not have the same elements in equal proportions.

95. From the facts stated in reference to roots, the importance of **deep, thorough** preparation of land before planting, and of shallow cultivation after the crop has been well started, is evident. Should the soil, however, become hard and baked, it would be good policy to cultivate deeply even at the risk of destroying rootlets, which under such circumstances are necessarily sickly, that new and vigorous ones may put forth and have mellow soil through which to penetrate. No fixed rule can be laid down to be rigidly followed in all cases. The intelligent farmer or planter should be guided by the indications of the season, and the condition of his crop, in the use of means necessary to meet an emergency. Even the bird, though governed by instinct, modifies

at times the shape of its nest to adapt it to peculiar surroundings, and so the farmer must use his **reason** in cultivating his crop. He should be guided, not by "cast-iron" rules, but by correct principles, and modify his mode of cultivation to suit each particular case.

CHAPTER VI.

THE IMPROVEMENT OF SOILS.

96. DEAD matter has no power in itself, so far as we know, to change into an organized structure. It has the power, under certain circumstances, of forming crystals of great beauty, but in these forms there is no **life**. This life-principle, every seed or germ has within itself, and when called into activity it unfolds the plant. Exactly what this life-principle is, we know not, but we can study the conditions of plant-growth, and explain to some extent the changes which take place in the swelling of the seed by heat and moisture, and its partial decomposition into simpler forms, whereby energy is imparted in some mysterious way, sufficient to push the rootlet downward, and the slender stem upward to the sunlight.

97. The connection between the rays of the sun drunk in by the leaves of plants and the absorption of food by their roots, the appropriation of this food and its assimilation or use in the production of bark, and wood, and leaves, and fruit, has not yet been ex-

plained, and perhaps never will be. Chemistry, however, has taught us how to stimulate the process, and has shown us that there is no change of one element into another, but a simple use of that which is in reach of the plant, whether it be in the air above or in the soil beneath.

98. If it be absurd to suppose, as everybody must admit, that one element can change into another, the practical conclusion follows that, to insure fertility, the necessary elements which compose the particular plant under cultivation must exist in the soil, or its growth is impossible. As simple and self-evident as this statement may appear, there has ever been a vague notion in the minds of men that nature has some power of changing elements, or of supplying them as needed for the growth of vegetation.

99. It has not been many years since writers on agriculture affirmed that the inorganic constituents of a good soil are inexhaustible. Such statements are contrary to the teachings of science and of practical experience. It is true, the application of scientific principles in farming will not in every case produce a good crop, because there are conditions that human agency cannot control. The wind, and rain, and sunshine are given or withheld by a power which we cannot govern, and without whose favoring influence all our labor is in vain.

100. Science not only determines the kind and quality of food which plants need, but points out the localities where this food can be obtained, and fur

nishes the farmer with methods for preparing, in concentrated form, the very elements which are deficient
in an unproductive soil. Through the influence
of its teachings, the manufacture of fertilizers has
become one of the great industries of the world.
Substances in the highest degree offensive, injuri-
ous to health, and difficult to be gotten rid of,
have been converted into merchandise of great
value.

101. The disposal of refuse matter, by putting it
to profitable use, is one of the greatest results of
modern science. A portion at least of the sewage of
London, and other large cities of the world, is now
rendered harmless, and inoffensive, and even valuable as an article of trade. A perfect system of agriculture requires that it should *all* be returned to the
soil whence it came, and serve as food for growing
crops. The indestructibility of matter, in connection
with the correlation and conservation of energy, pre-
sents nature as a grand system of mutually dependent parts which move in a ceaseless round of universal harmony. It is for science to study the relation
of these parts, however vast or minute, and teach us
how to control their movements so as to promote
the interests of the human race. Much has been
done in this direction, while much remains to be
accomplished.

102. All truly scientific methods of improvement
of soils are founded on observation and experiment,
and are addressed to the reason and common sense

of intelligent men. If a soil has **all** the constituents
that plants need, and in the best possible condition
for use as plant-food, then such soil cannot be improved. If, however, one or more constituents are
deficient, or entirely absent, or not in the best condition to serve as plant-food, then improvement is
not only possible, but desirable.

103. The means used for the improvement of
soils are:

Mechanical, as draining, plowing, sub-soiling,
mixing with clay, etc.

Chemical, as manuring or fertilizing.

The first changes the physical properties; the
second, the composition. When necessary constituents are absent or deficient, no amount of draining
or sub-soiling will secure a good crop.

104. The extensive and thorough system of draining by means of **tiles** or **clay pipes**, which is used
in Europe and in some portions of our own country,
is too costly where land is cheap and abundant. In
such cases open ditches should be used to carry off
surplus water. Intelligent farmers understand very
well the importance of removing excess of water by
some sort of drainage that will be least likely to remove the soil with it, but unfortunately they do not
always put their knowledge into practice. Where
loose rock is convenient, covered ditches with ten or
twelve inches of rock at the bottom are easily made,
and form excellent drains. Deep plowing and sub-
soiling are means of draining, to a limited extent,

but cannot be substituted for ditching in wet, swampy lands.

105. The advantages of drainage are numerous, of which may be mentioned the following:

(1.) It makes the soil warmer. The evaporation of water from any surface is cooling. The heat of the sun falling on a wet soil is taken up in converting the water into vapor, and does not reach the roots of the plant.

(2.) It keeps the plant food from becoming too much diluted, and leaves it in a concentrated form for absorption by the plant.

(3.) It gives free access of air to the roots of plants. Water not only keeps out the air, but drowns or destroys the soil roots.

(4.) It aids in bringing about a proper decomposition of organic matter, and preventing the formation of organic acids that are hurtful. Swampy lands which are generally sour and unproductive for valuable crops are rendered productive by draining.

106. Deep plowing and sub-soiling not only assist in draining the soil, but also render it better able to stand a drought. The surplus water of heavy rains can sink down without washing, while the capillary action of a thoroughly pulverized soil draws moisture from a greater depth in time of drought, and the roots penetrate more easily beyond the influence of the sun's rays and find abundant food, while air can also reach a lower depth more readily and exercise its disintegrating effect.

4

107. Deep plowing also brings new earth to the surface, and forms a deeper soil, altering its physical and chemical properties and promoting the growth of vegetation. Should the sub-soil be less fertile than the surface, which is generally the case, care should be taken not to bring much of it to the surface during any one season. After the soil has been deepened sufficiently, the sub-soil should only be stirred by means of the sub-soil plow.

108. A very economical and effective method of sub-soiling is to run a plow with a long narrow shovel immediately behind the common turn-plow. The shovel can be made by a common blacksmith, and be easily repaired. The expense of sub-soiling can thus be made very light, while the advantages, especially in heavy soils, are sure to be recognized after a thorough trial.

109. There are other mechanical means of improvement, such as the mixing of stiff clays with light, sandy soils, and the opposite, which are practised with advantage in thickly populated countries, but which cannot be profitably employed by us in the cultivation of wheat, corn, cotton, and tobacco. The main object is to secure the greatest amount of improvement at the least expense. An unsettled feeling frequently prevails among farmers and planters in a new country which operates strongly against anything like permanent improvement, and tends to produce careless cultivation and rapid exhaustion of the soil. The abundant supply of new

lands in the far West tends to produce the same re.
sult.

110. It is a narrow-sighted policy that opposes
thorough cultivation and the use of fertilizers, on
the ground that this may be the last crop, and such
investments of labor and capital can not be remu-
nerative the first year. Thorough cultivation is
always proper. Whether fertilizers should be used,
as well as what kinds, depends upon circumstances,
which will be discussed in the next chapter.

CHAPTER VII.

THE USE OF MANURES, OR FERTILIZERS.

111. THE chemical improvement of soils embraces the use of fertilizers, or manures, the application of which depends on the following well-estab-
lished principles :

1. Plants derive an essential part of their food
from the soil. This includes all the inorganic elements found in the ash, and a variable quantity of
those organic elements which volatilize when the
plant is burned.

2. Different plants require a special supply of
different kinds of inorganic food, or the same kinds
in different proportions, which must be contained in
the soil.

3. Some soils have a deficiency of plant-food

which must be supplied before particular plants can grow. The food may be in the soil, but not in a condition to be appropriated by the plant.

112. Soils may be naturally deficient in important constituents, or they may have become so by long-continued cultivation and removal of crops. In such cases, improvement is effected by the addition of a suitable fertilizer, or manure.

113. The continued removal of crops is sure to produce exhaustion. The following table shows the quantity and composition of the ash contained in one English ton (2,240 pounds) of hay of different kinds, and which is carried off when the hay is taken from the farm (the numbers represent pounds):

| | Italian rye-grass hay. | Clover hay. | | Lucerne hay. |
		Red.	White.	
Potash............	17	26	24¾	30
Soda................	7	3½	10½	13½
Lime..............	13¾	55½	45½	107½
Magnesia	3	17½	14	7¾
Oxide of iron..........	1	1½	3½	⅔
Sulphuric acid.........	4	6½	12½	9
Phosphoric acid.......	8¾	10	20	29
Chlorine...............	2	4	5	6¾
Silica.................	81½	5	6	7⅓
	138	129½	141¾	211½

114. Animal products also remove valuable constituents, and in large quantities, as can be seen from the following table, which is based upon the experience of Messrs. Lawes and Gilbert, two noted English agriculturists :

Composition of Animal Exports from a Farm.

	Nitrogen.	Phosphoric acid.	Potash.	Lime.	Magnesia.
	Lbs.	Lbs.	Lbs.	Lbs.	Lbs.
Fat ox, per 1,000 lbs. fasted live weight....	23.13	16.52	1.84	19.20	0.63
Fat sheep, per 1,000 lbs. fasted live weight..............	19.60	11.29	1.59	12.80	0.53
Fat pig, per 1,000 lbs. fasted live weight	17.57	6.92	1.48	6.67	0.35
Milk, 1,000 lbs............. ...	5.25	2.03	1.80	1.56	0.16
Wool, unwashed, 1,000 lbs.....	73.00	1.00	40.00	1.00	0.70

By "fasted" is meant that the animals were killed when their digestive organs were empty, or nearly so.

115. The rapidity of exhaustion will depend upon the kind of products removed. Some writers assert that the amount of plant-food in good soils is so great, that practically they can not be exhausted, but the experience of farmers proves the contrary. Good crops of wheat have been grown on the same soil for a number of years without manures, but careful experiments show that there is a gradual falling off in the yield.

116. The fertilizers, or manures used for improving soils and restoring fertility, may be divided into three kinds, **vegetable, animal,** and **mineral.**

117. As every decaying plant contains all the constituents necessary for the growth of a similar plant, of course it furnishes a good fertilizer. Vegetable substances, such as clover, pea-vines, etc., are sometimes plowed in with great benefit to the soil. They not only furnish material properly prepared as

available food, both organic and inorganic, but they are also highly useful in making stiff lands mellow and porous, and light soils retentive of moisture.

118. Cornstalks laid in a furrow and the earth bedded upon them, and even weeds when turned under, produce a fine mechanical effect on stiff lands apart from their value as chemical manures. Whether green or dry, these vegetable substances soon undergo decomposition, and give up their constituents as food to growing plants.

119. Cotton-seed is one of the best and most energetic of this class of fertilizers. It contains a great deal of nitrogen, which has a wonderful effect in stimulating vegetable growth. Leaves, and all sorts of vegetable matter, can be profitably used, which, as a top-dressing, produce warmth, and as they decay furnish valuable plant-food.

120. The large amount of cotton-seed produced in the South has caused special attention to be called of late years to its value and to the best method of using it as a fertilizer. Like other seeds it is rich in potash and phosphoric acid, two important constituents of a fertile soil, which are generally in such limited quantities that in the process of removal of crops they are the first to become deficient. Cotton lint contains only about 1 per cent. of ash or mineral matter, while the seed contains about 9 per cent. In a bale of cotton, the yield of say 1,700 pounds of seed-cotton, we have only five pounds of mineral matter in the lint, and one hundred and eight in the seed.

The removal of the seed, therefore, is more than twenty-one times as exhaustive of mineral matter as the removal of the lint.   .

121. If the seed be returned to the soil, there is no marketable crop less exhaustive of mineral matter than cotton, but if the seed be permanently removed the result is quite different. In view of the importance of restoring this valuable substance to the soil, the question arises, How can it be done with the greatest advantage to succeeding crops? Like all organic substances it must undergo decomposition before it can serve as plant-food. If this decomposition be effected in a heap without mixing with earthy matter, the ammonia generated will mostly escape. This escaping ammonia is the most energetic stimulant known, and should, by all means, be preserved.

122. An economical mode of application is to grind or crush the seed, scatter it in deep furrows, and cover it up some time before planting. The practice of scattering the seed in the drill at the time of planting does very well, provided too large a quantity is not used. The decay of cotton-seed is a species of combustion, or fermentation which generates considerable heat, and if the roots of the young cotton-plant are in contact with the hot, fermenting mass, they are almost sure to be destroyed. If, however, the fermentation takes place below the plant, with a layer of earth between, the young plant will be stimulated and nourished by the escaping vol-

atile products of decomposition, and the roots thus strengthened will reach down, and appropriate the food contained in the decaying seed.

123. To those living near an oil-mill, there will be great economy in having the oil first extracted from the seed, and then using the oil-cake as a fertilizer. The oil is a product of great value for certain purposes, but of little value as a manure, because it contains only the organic elements, carbon, hydrogen, and oxygen, which are supplied by the atmosphere to every plant.

124. The process of grinding or crushing the seed promotes decomposition, and hastens its availability as plant-food; but the expense of the necessary machinery is too great for general use, unless the kernel is to be used as food for stock, for which purpose it is admirably suited. The valuable products of decomposing cotton-seed may be preserved by composting, or mixing with other fertilizers which can absorb or combine with its volatile products.

125. The important object to be accomplished, on every farm or plantation, is the restoration to the soil, in some form or other, of the large amount of valuable mineral matter contained in the seed, and the prevention of its permanent removal to the gradual exhaustion of the land. In Europe, rape-seed and linseed-cake, similar to our cotton-seed oil-cake, are highly esteemed as fertilizers.

126. **Animal** fertilizers, consisting of the flesh of all dead animals, with the scrapings of tanneries and

the offal of slaughter-houses, are among the most stimulating substances used to promote the growth of vegetation. The animal structure contains every element of the plant except **silicon**, and some portions of the body contain most valuable constituents in a highly concentrated form. The products of decomposition are especially rich in nitrogen.

127. The bones and excrements of animals, among the most powerful and valuable of fertilizers, generally undergo some manipulation, or preparation before being used, which renders them more suitable for immediate use, or more convenient for handling.

128. Raw bones when dried consist chiefly of tricalcium phosphate (common phosphate of lime) and gelatinous matter, in the proportion of about two-thirds of the former to one-third of the latter. Their average composition may be stated as follows:

	Per cent.
Animal matter............................	33
Tricalcium phosphate..	57
Calcium carbonate.......................	8
Calcium fluoride.........................	1
Magnesium phosphate....................	1
	100

129. Bones are sometimes merely ground to a fine powder, and sown broadcast, or drilled with the grain at the time of planting. In this case their action is slow, because they are insoluble in water, and

their constituents must be rendered soluble before they can be taken up by the plant and used as food. This change will take place gradually in the soil.

130. In order to hasten the action of bones, it is customary to convert them into what is known as **superphosphate.** This is done by mixing .the broken or ground bones with about half their weight of common sulphuric acid, or oil of vitriol diluted with two or three times its weight of water. After the acid has acted upon the bones, a somewhat pasty mass is formed, which may be mixed with ashes, sawdust, rich earth, refuse of salt-refineries, etc., to neutralize excess of acid.

131. The insoluble bone, or calcium phosphate, is thus changed by the acid into a soluble phosphate, commonly known as **superphosphate.** The sulphuric acid removes a portion of the calcium from the bones forming calcium sulphate, or gypsum, which is itself an excellent fertilizer. The addition of ashes to neutralize excess of acid is an advantage to the mixture, but unfortunately manufacturers or dealers sometimes add sand and earth of no value, in such quantities that suspicion has been cast upon the genuineness of much of our commercial superphosphate.

132. A large deposit of tricalcium, or bone phosphate, the remains of extinct animals, was discovered near Charleston, South Carolina, some years ago. It furnishes an abundant supply of this material, and has caused a reduction in the cost of superphos-

phates. An analysis of South Carolina phosphate shows it to contain ·

	Per cent.
Moisture	7.79
Organic matter and water of combination	4.60
Silica	10.35
Calcium carbonate	8.20
Tricalcium phosphate	61.89
Earthy and alkaline salts	7.17
Total	100.00

133. Hundreds of tons of cattle and buffalo bones are brought from the Western Plains, and are either used in the form of ground bones, or first converted into superphosphate. The following mean of many analyses of superphosphates, taken from " Phosphates of Commerce," by Jones, will show their general composition :

	Per cent.
Moisture	14.50
Organic and volatile matter	12.91
Monocalcium phosphate, or superphosphate *.	17.30
Tricalcium, or bone phosphate	4.16
Calcium sulphate	47.78
Alkaline salts	0.15
Sand	3.20
Total	100.00

* Equal to 27.08 per cent. of bone phosphate rendered soluble. "Monocalcium phosphate," in the above statement, is sometimes called "biphosphate of lime." It is, according to the modern chemical nomenclature, tetrahydrogen calcium di phosphate.

134. The manufacture of superphosphates is perhaps too difficult for the ordinary farmer, and yet in some cases it can be made profitable. To any one disposed to try the experiment, the following directions are given, which, if strictly followed, will insure a good result :

Secure a carboy of sulphuric acid, commonly known as "oil of vitriol," break up the bones as fine as possible, and to one hundred pounds of bones add forty pounds of acid, previously mixed with twice its bulk of water, and cooled. The mixing of the acid and water together produces great heat. It is best to pour the acid into the water with constant stirring. After the mass has become pasty by solution of the bones, add ashes, mix thoroughly, and allow the mass to dry. It can easily be reduced to a powder, and thus applied to the soil. A long trough or one or more barrels may be used in making the solution, or the finely broken bones may be placed in a pile on a suitable floor, and sprinkled from time to time with diluted acid. The acid must be handled with care, as it is highly corrosive. From two to three hundred pounds of this mixture should be used per acre.

135. Dr. Nichols, in his " barn-floor lecture," gives the following practical directions for making superphosphate from charred bones, a substance used by sugar-refiners for decolorizing syrups, and then sold to makers of fertilizers :

" A box four feet square and one foot deep is

lined with thick sheet-lead—the lead in one piece, soldered at the corners strongly with **lead** solder. Its capacity is just right for making one-fourth of a ton of superphosphate at a time, and it requires a whole carboy of vitriol, so that no fractional parts of acid are left to cause trouble. It requires :

One carboy of oil of vitriol.... .... 165 pounds
Bone-charcoal 380   "
Water............................... .... 10 gallons

" The water is first placed in the trough, and the acid is added to it gradually, causing a great boiling, with evolution of heat and steam. It takes about an hour for the reaction to become complete, and then it will soon dry and be free from moisture. It needs no grinding, it is ready for the field as soon as cool."

136. **Guano,** one of the most costly and valuable of fertilizers, was in use by the inhabitants of South America, when the Spaniards under Pizarro overran and subdued the powerful empire of the Incas. Prescott, in his history of " The Conquest of Peru," gives an interesting account of the progress made in agriculture by the aborigines of that country. In some places where the dry sandy plains were unproductive, they sank immense pits, some of which were an acre or more in extent, and dug out to a depth of fifteen or twenty feet, in order that moisture sufficient for vegetation might be afforded. Other conditions of fertility were supplied by thorough pul-

verization of the soil and a systematic application of fertilizers.

137. The fertilizer in common use among the Peruvians was guano, the excrement of immense flocks of sea-birds which frequent the rocky islands near the coast. Countless numbers of birds have for centuries hatched and reared their young on these islands, leaving deposits which cover a large extent of surface, and are in places from twenty to fifty feet in thickness. It was estimated, thirty years ago, that on the Chincha group of islands there were from twenty to twenty-five millions of tons. It has been so long accumulating that in some localities the deposit is found in layers with sand like a regular geological formation.

138. The importance of this fertilizer was so highly appreciated by the old Peruvians that, when the Spaniards first visited the country, special laws were in operation to regulate its use, and enforce its application to the soil by those engaged in agriculture. Its value was not known or appreciated in Europe or in our own country until within the last thirty or forty years. The credit of introducing it into Europe is ascribed to Humboldt. The first cargo was taken to England in 1826, but its value was not understood until a number of years afterward.

139. Liebig, in his "Letters on Modern Agriculture," says : "Before 1840, guano had never been used as a manure on a European field. When the first

vessel loaded with guano arrived at Liverpool, numerous experiments were made with the new manure which proved failures, and agriculturists were not agreed about its utility until they had practically tested its use. Since that time, many hundreds of ships have passed to and fro, and have brought to the European Continent guano to the value of about 300,000,000 florins (about $125,000,000) ; and, within the same period, there has been produced a surplus of more than 400,000,000 cwt. of corn, or of its equivalent in flesh. It is true, guano would have found its way to Europe even without the recommendation of science, but it would not have made its way so speedily. In the late period of sterility through which we have passed, it has been the means of alleviating the wants of many millions of men."

140. The first shipment of guano to the United States was made in 1845, and up to June 30, 1855, about half a million tons had been imported, while the amount sent to Europe during the same period was vastly greater. The Government of Peru has derived a large revenue from the sale of this valuable fertilizer, and the ships of all nations have engaged so actively in the trade that the supply on the islands has greatly diminished. The ocean has been searched for other islands containing similar deposits, and, while some have been found, none are esteemed so highly as those on the coast of South America.

141. .On these Peruvian islands it seldom or never rains, while the hot sun rapidly evaporates the moisture naturally contained in the deposit, and leaves the valuable constituents in their most concentrated form. On other islands frequented by sea-birds, a great deal of that which is valuable in the deposits is washed out by rains more or less frequent. Guano has been brought from a few islands on the coast of Africa and other parts of the world where rain seldom falls, but none has been found equal in value to the Peruvian.

142. The guano-producing birds feed on fish, a food especially rich in phosphates and nitrogenous compounds. The mixture of their solid and liquid excrements contains in a highly concentrated form those very elements which are most likely to be deficient in soils, and hence its universal applicability.

143. The composition of guano varies considerably in different samples. An average analysis of Peruvian guano imported into Ireland in 1876 is given by Johnston and Cameron as follows:

	Per cent.
Moisture	13.13
Organic matter, etc.*	48.17
Calcium phosphate	26.58
Alkaline salts	11.00
Insoluble matter	1.12
	100.00

* Yielding ammonia................... 11.80.

144. Peruvian guano contains less moisture than African and Patagonian because the climate of Peru is drier. It also has a less pungent smell of ammonia. As guano is often adulterated with sand, earth, etc., the following physical properties and tests will enable one who is not a chemist to distinguish between a good guano and one of inferior quality, or one which has been adulterated.

145. Genuine guano is a substance of a yellowish-brown color, of a peculiar urinous odor, and has mixed with it, white lumps or fragments. When heated on a shovel or iron plate, at least half of it will volatilize, and nearly the whole of the remainder will dissolve in dilute hydrochloric acid. Sand and earthy adulterations will be left undissolved, which in good guano does not amount to more than from 1 to 2 per cent. When first heated, it gives off white vapors with a strong smell of ammonia. This odor is greatly increased by the addition of lime. The ash should be white or grayish; a yellow or red ash indicates an admixture of clay or earthy matter.

146. The highly stimulating effect of guano is due to the large amount of nitrogenous matter which it contains. The atmosphere furnishes nitrogen to plants in the form of ammonia, but not in sufficient quantity to meet the demands of a luxuriant growth of wheat, corn, or other crop. This must be supplied by decaying nitrogenous matter, salts of ammonia, or nitrates.

147. The newly sprouted plant stands ready to grow, it may be with all its organs perfect, and with every article of food at hand, but, so long as there is only a limited supply of one important element, the rapidity and vigor of its growth will be retarded. The air contains an abundant supply of nitrogen, but not one atom of it, as chemists believe, is contributed directly to the growth of plants. This important element, before it can be used as plant-food, must be in union with some other element, and guano furnishes it in the best possible combination. The phosphates, too, which exist so largely in guano and other salts, give additional value to this fertilizer.

148. Guano, then, as we have shown, promotes the growth of vegetation by furnishing a large supply of nitrogen in an available form, and a few other highly important constituents of our most valuable crops. As it does not contain all the elements of plant-food, there may be soils which it will not greatly benefit unless mixed with other substances. It is well to keep in mind the fact that **all** the elements of plants are necessary for growth. The reports of failure in the use of guano and other fertilizers, and statements to the effect that they "run out" and, in the end, "make the soil poorer," may be explained by the fact that such manures lack some elements which every fertile soil must contain ; or they may so stimulate a comparatively poor soil that in a few years the supply of some important element which the guano or other manure lacks will be exhausted.

Of course, a further application of guano without the addition of this deficient element will prove a failure. The fault is not in the fertilizer, but in not supplying other material.

149. It is customary to mix guano with other fertilizers not so rich in ammonia, and thus have a manure that can be used profitably for almost any crop. A large number of "manipulated guanos" are sold, and where the manipulation, or mixing with other fertilizers, has been fairly and honestly done, such mixtures are highly beneficial and economical.

150. Peruvian guano and other strongly nitrogenous manures should never be placed in immediate contact with the seed, because the ammonia given off is very caustic, and will kill the seed-germ. It may be placed under the seed with a few inches of earth between, or at the side, the former being preferable.

151. A mixture of one hundred pounds of Peruvian guano with two hundred of a good superphosphate per acre gives excellent results. The ingredients should be thoroughly mixed, sown broadcast, and turned in with the wheat or other grain, or they may be placed under the bed on which cotton or corn is to be planted. In wheat-growing States, drills are used both for planting the wheat and for depositing the guano and superphosphate. Some farmers prefer scattering the fertilizer broadcast, whether for cotton, wheat, or other crop.

152. It will not do to mix guano with fresh ashes

or lime, because the potash contained in ashes, as
well as lime, sets free the ammonia.   This ammonia,
in all ammoniacal fertilizers, is either free or in combination with acids.   Potash, soda, and lime have
a stronger affinity for these acids than the ammonia,
and hence they take its place.   The ammonia being
a gas, passes off and diffuses itself in the air.   It is
important to understand the chemical principles in-
volved in mixing fertilizers, as well as in their action
when applied to the soil.

153. **Stable** or **farm-yard** manure is the most
common of fertilizers, and one which every farmer
and planter can have in great abundance without
much trouble or outlay of money.   It only requires
the exercise of a little care in its preservation in order to secure a large supply on every farm.   Stable-
manure consists of the solid and liquid excrements
of animals which feed on some of our most valuable
crops, mixed with portions of these crops in a more
or less advanced stage of decomposition.   It constitutes a stimulating and nutritive fertilizer applicable
to all crops.

154. The value of stable-manure depends to some
extent on the food of the animal; the richer the food,
the better the manure.   The liquid portion is especially rich in phosphates and compounds of nitrogen,
and should never be allowed to run to waste.   Whenever a manure-heap gives out a strong smell, some
of its valuable constituents are escaping.   In such
cases, the smell of ammonia may not be perceptible,

but its escape can easily be determined by the common test for this substance, that is, the formation of a white vapor with hydrochloric acid. A glass rod dipped into this acid, and brought in contact with the gaseous ammonia, will show the effect mentioned. To prevent the escape of ammonia, the heap as it accumulates should be covered or mixed with earth, or sprinkled occasionally with dilute hydrochloric, or sulphuric acid, or solution of ferrous sulphate, known as copperas. Lime and ashes must not be added, because they would release the ammonia, instead of retaining it.

155. Manure-heaps should be kept moist in order that fermentation or decomposition may go on, but should not be exposed to violent rains which wash out soluble ingredients of great value. Perhaps the best and most economical plan for general use, is to pile the manure carefully in rail pens and cover lightly with boards, and, even if not covered, the pens will prevent the scattering and wasting of the manure. The excrements of all our domestic animals should be added to these piles, together with the sweepings of yards, and refuse of all kinds. By this means, with a little attention and no outlay of money, much valuable material which is now lost could be restored to the soil.

156. Human excrements, so abundant in large cities, form a manure of considerable value. When dried, powdered, and mixed with charcoal, gypsum, etc., it is sold under the name of **poudrette.** Large

quantities are prepared in Europe which command a good price, but in this country it is used to a very limited extent. The efforts thus far made to utilize the sewage of cities have in some instances proved successful, but there are difficulties yet to be removed before its profitable use can become general.

157. **Animal manures** are richer than vegetable because they contain more nitrogen, which in the process of fermentation and decay unites with hydrogen to form ammonia, and because they also contain a larger proportion of valuable inorganic elements. The effect of respiration and digestion, the two great vital functions of the animal system, is to extract carbon and hydrogen from the food through the blood and lungs where they escape as carbon dioxide and water. The water in large quantities passes through the pores of the skin, leaving the other constituents in a more concentrated form.

CHAPTER VIII.

MINERAL FERTILIZERS.

158. THE list of **mineral fertilizers** embraces a large number of substances, only a few of which, however, will be mentioned, such as lime, marl, gypsum, salt, and ashes.

159. **Lime,** or calcium oxide, is made by heating common limestone in a suitable kiln. This mineral, when pure, consists of lime in combination with carbon dioxide, or carbonic-acid gas. The rock is broken into small-sized pieces, placed loosely in what is known as a **kiln,** and, by means of wood or coal, thoroughly heated until the carbon dioxide is driven off. Calcium oxide, or quicklime, is left. One hundred pounds of pure limestone yield fifty-six pounds of lime.

160. If water be poured on quicklime, a portion of the water will combine with it, giving out heat, and forming **slaked** lime. The same result will be produced by exposing the quicklime to moist air. Long exposure to air will result in a reunion of carbon dioxide with the lime. The new carbonate will be in a finely divided condition, which is better for fertilizing purposes than the original limestone, though not so good as the slaked lime (30).

161. One hundred pounds of quicklime (also called caustic lime) unite with thirty-two pounds of water in the process of slaking. This water is not mixed with the lime, but forms a chemical compound known as **calcium hydroxide.** This hydroxide is slighty soluble in water, of caustic taste, and when mixed with sand forms common **mortar.** The hardening of mortar is due to the absorption of carbon dioxide from the air, and a gradual union of the lime with the silica, or sand.

162. The application of lime to soils furnishes at

least one mineral constituent to plants, one, however, that generally exists in soils, and hence its valuable and often surprising effects are not to be attributed to the plant-food it contains so much as to other important purposes which it serves. Its alkaline properties cause it to neutralize acids, which sometimes exist injuriously in soils. It also renders stiff clays light and mellow, and aids in the decomposition of organic substances, and of some insoluble inorganic compounds. In other words, lime corrects "sourness" in lands, destroys excess of vegetable matter, lightens heavy clay soils, and releases potash for the use of plants by decomposing silicates.

163. **Marl** is a mixture of calcium carbonate—derived chiefly from the shells of animals—clay and sand in variable proportions. It is generally valued according to the amount of calcium carbonate it contains, which may vary from five to ninety per cent., though in almost all marls there are other valuable constituents. The calcium carbonate, or carbonate of lime, as it is generally called, is usually in a finely divided state, and can be readily used as food by the plant.

164. The celebrated New Jersey green-sand marl contains a large percentage of potash, which gives it great value. Excellent marls are found in many parts of the country, and need no preparation before using. The following analysis by Professor Norton will give some idea of the composition of an excellent marl ·

	Per cent.
Calcium oxide (lime)	35.00
Carbon dioxide	45.02
Oxide of iron and aluminium, with traces of phosphoric acid	2.69
Magnesium oxide	0.66
Organic matter	7.06
Sand	9.57
	100.00

165. Some marls consist almost entirely of shells more or less broken; in others, clay or sand predominates. Good marl always gives off a large quantity of carbon dioxide on the addition of a strong acid, and in this way some estimate may be formed of its value (31). This fertilizer is particularly well suited for application to sandy lands, supplying the very elements which such lands generally need for the continued production of good wheat and cotton crops.

166. Gypsum, or calcium sulphate, also called land plaster, is a very important mineral fertilizer and has long been in use as a valuble manure for corn, tobacco, wheat, clover, and the grasses generally. It is composed of lime and an oxide of sulphur in chemical combination with a large percentage of water. This water can be driven off by heat leaving a white substance which, ground to powder, forms what is known as plaster of Paris. In this form, it has the property of recombining chemically with water and forming a substance of the same composition as the

original gypsum. This property fits it admirably for taking casts of statuary, for forming the "hard finish" on the plastered walls of our houses, and for stucco-work of all kinds. Dentists use it in large quantities for taking casts of the mouth, and for other purposes in their work, while it is useful to artists in forming models from which to copy and perfect the creations of their genius (32).

167. Gypsum is found in nature as a mineral, generally as a white compact mass, though sometimes in a transparent crystalline condition, as in the mineral selenite. Alabaster is also a form in which this mineral is found, and used in the manufacture of vases and other ornamental work. The rock is quite soft, easily scratched with a knife, and whitens on being heated. It contains about twenty-one per cent. of water. As a fertilizer, it furnishes lime and sulphur to plants, and is thought to have the power of absorbing ammonia from the air and supplying it to the plant. To this important property Liebig ascribes much of its wonderful effect upon young grass and wheat. It is usual to sow it broadcast over wheat and grass during the early spring, and to drop it with corn at the time of planting, or to drop a small portion on each hill of corn after the corn has been thinned. Corn moistened with water and rolled in plaster at the time of planting, will get a more vigorous start, and be the better enabled to stand an early drought. Cotton-seed may also be very advantageously rolled in plaster previous

to planting. For this purpose, the seed is placed in a tub of convenient size, thoroughly moistened with water, the plaster added, and the whole well stirred. As much as will adhere is dropped with the seed.

168. Gypsum, or plaster, has no caustic properties like quicklime and guano, and therefore seeds are not injured by being placed in immediate contact with it. The vigorous, healthy start which it gives to the young plant is very desirable for both corn and cotton, since weak, sickly plants are almost sure to suffer from insects, or perish from other causes. The noted English farmers, Lawes and Gilbert, found, during four years, an average increase of nearly one ton of hay per acre by the use of gypsum as compared with adjoining land without gypsum. This is one of the cheapest of fertilizers, and should be used by farmers and planters who desire an increase of production by a moderate outlay of money. It is not necessary to mix gypsum with superphosphate, because, in the process of making superphosphate from bones, calcium sulphate, or gypsum, is always formed at the same time, and constitutes a large part of the superphosphate.

169. **Common salt** has been used as a manure from very ancient times. It can be used with advantage only in small quantities, as a heavy application destroys vegetation. Common salt, or sodium chloride, consists of two elements, chlorine and sodium, and is a valuable addition to compost-heaps.

Its value as a manure is denied by some experiment-
ers.   It acts, no doubt, in an indirect manner, releas-
ing nitrogen from some of its compounds and aiding
in the solution of calcium phosphate and other salts.
According to some writers, salt may be used to advantage for checking growth on lands which produce such an excess of straw as to cause wheat to
fall before the seed is matured.

170. In chemical language, **salts** are compounds
in which the hydrogen of an acid is replaced by a
metal.   The union of a non-metallic element with a
metal forms a salt like sodium chloride.   Another
class of salts have oxygen in them in addition to the
other non-metallic element, like potassium nitrate or
nitre, sodium nitrate or Chili saltpetre.   These are
used as fertilizers, as are also sulphates of various
kinds, and the refuse **salts** from salt-works and
chemical manufactories.   The utilization of what
were formerly **waste products** is carried to such
an extent in Europe that scarcely anything is lost.
Every product has a marketable value, and what
cannot be consumed by animals, or put to some
useful purpose, is returned to the soil as food for
plants.

171. Among mineral fertilizers may be classed
the **ash** of vegetable substances whether of wood,
coal, or plants.   The following result of the analysis
of the ash of oak and beech wood shows the nature
of this fertilizer:

Percentage of	Oak.	Beech.
Potassium oxide..................	8.43	15.83
Sodium oxide.......................	5.64	2.79
Sodium chloride..................	0.02	0.23
Calcium oxide, or lime	74.63	62.37
Calcium sulphate...................	1.98	2.31
Magnesium oxide, or magnesia........	4.49	11.29
Iron oxide, or ferric oxide.	0.57	0.79
Phosphoric pentoxide or phosphoric acid.	3.46	3.07
Silica...............................	0.78	1.32
	100.00	100.00

172. The ash contains all the mineral matter which the plant derives from the soil, and is in a good condition to be used again as plant-food. The potassium and sodium salts are easily washed out by rains, being very soluble in water. In the process of making soap, ashes are leached for the purpose of obtaining these two alkaline metals. Other substances, however, remain, so that even leached ashes are valuable as a manure.

173. A **compost** is simply a mixture of different fertilizing materials. They are generally heaped together and allowed to ferment, or decompose. Organic matter, such as straw, chips, cotton-seed, dead animals, refuse from slaughter-houses, etc., must undergo decay or putrefaction before they become valuable as manure, and in doing so they give off ammonia which will be lost unless they be mixed with earthy or other inorganic matter to absorb the gas as fast as formed. Composting is the art of so

mixing these that a valuable fertilizer will be formed, and no important constituents escape. Such mixtures of raw material must be kept moist, but must not be exposed to rains that will dissolve out soluble salts. On every farm or plantation properly managed, all such substances will be preserved and returned to the soil.

174. Pendleton, in his valuable work on "Scientific Agriculture," gives good directions for such a compost-heap, as follows: "A layer of stable-manure six inches thick, with a good sprinkling of ground phosphate over it; then a layer of cotton-seed three inches thick (previously saturated with water), and then another sprinkling of superphosphate, say half an inch thick; then a layer of stable-manure, and so on until the heap is completed, which should be conical in form. Over the whole heap, when sufficiently large, apply several inches of dry clay soil, if you choose, which will absorb every particle of the escaping ammonia. If, however, this crust should become so saturated as to allow it to escape, an additional coating of soil can be applied."

175. If the cotton-seed and superphosphate are not on hand, use, with the stable-manure, the sweepings of the yard, old mortar, leached ashes, bones, scrapings from poultry-houses and yards, swamp muck, the earth from old ponds, and any and every kind of waste matter that is usually thrown away or hauled off that it may not become offensive. Such

piles should be covered with a thin layer of earth, to be added to whenever bad-smelling gases are found to escape.

176. Professor Ville, a noted French writer on fertilizers, calls a "complete manure" one that contains nitrogen, potassium, phosphorus, and lime, in suitable proportions to meet the demands of a given crop. These four elements are the only ones likely to be deficient; and he advises farmers to purchase the chemical salts containing these elements, and mix them according to formulæ given in his work on "Chemical Manures." A good superphosphate with nitre, or with some salt of ammonia, and one of potassium, would constitute such a mixture.

CHAPTER IX.

ROTATION OF CROPS.

177. THE fact has long been known that it is not best to grow the same kind of crop on the same land for a number of years in succession. Thus, the yield of corn, wheat, oats, tobacco, or other crop, if grown on the same land without change will gradually diminish, while if these crops be made to alternate, that is, first one and then another, the aggregate product will be much greater. This change is called **rotation of crops**.

178. The advantages of rotation, or change of

crops, result from the following considerations : In the first place, different crops require elements in different proportions ; one requires more potash than another, or lime, or phosphoric acid, or nitrogen, or some other constituent. By reference to the table in paragraph 54, it will be seen that potatoes require more potash than wheat or corn, while these require more phosphoric acid; clover and tobacco need a great deal of lime, and so with other crops. A number of crops of potatoes in succession, without the addition of potash in some form, will use up the available supply, unless the quantity be very large. Before this comes to pass, another crop that requires less of potash but more of some other element might grow well, while the potatoes would not flourish.

179. In the second place, rotation of crops gives time for the disintegrating action of the atmosphere, rain, and frost to prepare new material from the rock-particles in the soil, and get it in a form to be used by the plant. One crop may use up the available food of a particular kind faster than it can be prepared by these natural agencies.

180. In the third place, rotation, or change of crop, when properly managed, enables one plant to prepare food for another. Thus clover sends a long tap-root deep down into the soil, and brings up food to the surface. When the roots decay, the wheat-plant that has surface-roots mainly can use the food prepared by the clover.

181. In the fourth place, different crops require different modes of cultivation, so that the physical properties of the soil are improved by rotation. Grass-lands in a few years become hard and require to be loosened up, which can be done by the cultivation of a crop of corn, followed by one of wheat, and the re-sowing of grass

182. Not only is a change of crops desirable, but an occasional change of **seed** is found to be of great benefit. Wheat grown on stiff clay-lands for some years will be improved by getting seed from that grown on sandy soil, and that on sandy soil by obtaining seed from wheat grown on stiff clay. An occasional change of seed from one latitude to another is alo found to be beneficial.

183. Crops that require the same elements in about the same proportions should not follow each other, nor those that are similar in their mode of growth. Wheat and corn which depend mainly on surface-roots will do well after clover, cotton, or tobacco, which have long tap-roots that extend down into the subsoil. The greater the difference in the constitution and character of two crops, the more likely are they suited to follow each other. Climate and soil have much to do in determining the best rotation.

184. A good rotation for three years is—

	First year	Corn
	Second "	Wheat
6	Third "	Clover

A good four years' rotation is to allow the clover
to remain two years. In England, where corn is
not raised, a popular four years' rotation is—

First year	Turnips or other root-crop
Second "	Barley
Third "	Clover
Fourth "	Wheat

A six years' rotation :

Clover	Corn
Wheat, two years	Wheat
Clover	

185. For the tobacco-planter, a good rotation is—

First and second year Clover	
Third year	Tobacco
Fourth "	Wheat

This will keep the land fertile, or even improve
it. For the cotton-planter—

First year	Clover or peas
Second "	Cotton
Third "	Wheat

In much of the cotton-growing region in the
South, cotton is grown year after year without
change, to the great detriment of the land, except,
perhaps, in river-bottoms, where the soil is im-
mensely rich and very deep.

186. It is best in the above rotations not to remove the crop of clover, but let it lie on the ground,
and be turned under at the proper time. In fact,
there is no better way to improve land than by plow-

ing in the clover-crop after it has fallen on the ground, and undergone partial decay. The long tap-roots of this plant, as heretofore mentioned, go down very deep, and work up a great deal of material as food, and when the plant decays it furnishes all its elements to wheat, corn, or other crop that may follow. Old farmers know that, if land will only produce clover, it can be improved.

187. It was once thought that the advantages of rotation are due in some measure to the excretion, or giving out, of material by the roots of one plant which serves as food to another. It is not believed now that plants have the power to excrete, or give out, any such material. Plants do have some power of **selection** in taking in their food, for we know that, when different plants are grown in the same soil, they contain different proportions of the same elements, but there is no proof that they all take in the same substances and give out from their roots what each does not need for growth.

188. No kind of rotation will secure good crops if any one or more of the elements which a crop needs be entirely absent from the soil, or if the food is in such a condition that plants can not appropriate it. In such cases, fertilizing, or manuring, is the only way to restore such a soil or make it fertile. It is true that wheat may be grown for many years in succession on some qualities of land without manure, but experience shows that the crop gradually diminishes. The exhaustion may be slow, but it will surely come

CHAPTER X.

THE SELECTION AND CARE OF LIVE-STOCK.

189. EVERY farmer and planter finds it absolutely necessary to keep some kind of live-stock on his farm or plantation. Such work as plowing and hauling requires horses, mules, or oxen, while a variety of products raised on every farm can only be made profitable by being fed to cattle, sheep, and hogs, and thus turned into beef, mutton, pork, and bacon. The proper selection and management of such stock become matters of great importance. Attention will be called to a few guiding principles.

190. The kind of stock to be kept, above what is necessary for the absolute wants of the farm in the way of performing work, and furnishing food, will in each case depend upon the system of farming adopted as best suited to the nature and size of the farm. In selecting stock, whether sheep, cattle, horses, or hogs, the same common-sense principles should govern as in other things. **The best breeds**, that is, those best suited for particular purposes, should be selected, as they are in general the most economical. If the farmer wishes to raise beef, or mutton, or pork, for the market, he should select those breeds that can make the most flesh out of a given amount of food. Meat-producing animals are machines for converting vegetable food into flesh, and those breeds that give the greatest yield with the least care and

expense of management are the best. If the production of milk, or of wool, be the object in view, the same principle should govern.

191. In the management of live-stock of every sort, **kind treatment** is absolutely necessary to success. A poor, half-starved, ill-used horse or cow returns no profit to its owner, and is a disgrace to the farmer. The feeding should be **regular** and **uniform**, and proper shelter from rain and cold should be provided. Exposure retards the full, healthy development of young animals, and prevents the conversion of the food of older ones into salable products. As in the human system, so in the bodies of lower animals, the digestive process, to be successfully carried on, requires good food, pure air, and comfortable surroundings.

192. Plants derive their food from the soil and work up the earthy, or inorganic, material into organic products, such as sugar, starch, gum, oil, woody fibre, gluten; farm-animals derive their food from these organic products, and form therefrom fat, muscle, blood, and bones. The plant changes earthy matter, carbon dioxide, and ammonia, into organized products in which is stored up a great deal of force or energy; animals consume these organized products, which, in becoming disorganized in the body, give out force or energy to the animal in the form of **heat and animal power.** Plants store up energy to be used by animals.

193. The body of an animal is much more com-

plicated in its structure than that of a plant. In the higher animals it consists of a bony skeleton, covered with flesh, through which runs a network of nerves and blood-vessels. Within a portion of this skeleton are placed the lungs and digestive organs. Solid and liquid food passes through the mouth into the organs of digestion, where the portion suitable for conversion into flesh is made soluble by means of proper secretions. This nutritive portion is absorbed by suitable organs, carried into the blood, and then distributed to all parts of the body, where it becomes flesh and bone by a mysterious process which has never been explained. The operation is so delicate and complicated, and so many organs or parts are necessary to have it carried on successfully, that it is no wonder animals cannot thrive when badly treated.

194. It is well known that oxygen of the air is as necessary for animal life as for the burning of wood and coal in our fireplaces, and that air coming from the lungs in breathing is charged with carbon dioxide. This can easily be shown by breathing through clear lime-water by means of a glass or other tube. The water becomes milky, a result caused by the carbon dioxide from the lungs combining with the lime in solution to form insoluble calcium carbonate. Just as the chemical action, or burning of fuel in a fireplace or under a steam-boiler, produces heat and force, or energy, enough to move machinery, so the chemical action within the body

produces animal heat and muscular energy. In fact, no force, or energy, can be exerted by an animal without the destruction or consumption of material, and the heat of the body, like the heat of a fire, can not be maintained without the burning of fuel of some kind.

195. In cold weather, therefore, animals require more substantial food than in warm weather, in order to keep up a proper temperature of the body. Work-animals also require more food than those that do no work, in order to supply the necessary force, or energy. A certain amount of food is required to keep up the ordinary heat, and sustain the daily " wear and tear " of the body ; if the animal is expected to grow or become fat, an additional supply is necessary. All farmers know that hogs fatten faster, other things being equal, when kept in close pens, and not allowed to run at large. The physical exercise consumes much of the food that would otherwise go to form fat. On the same principle, the stall-feeding of cattle is economical.

196. Some years ago, Liebig, a noted German chemist, divided the food of animals into two kinds, **heat-forming** and **flesh-forming**. The first contains no nitrogen, like sugar, starch, and fat; the second contains nitrogen, like gluten and albumen. Liebig taught that the first kind is used altogether for keeping up the animal heat and forming fat, while the second supplies force, and forms flesh, or muscle. It is now believed that this theory is not

strictly true, as both kinds of food produce heat and energy. It is true that an animal cannot live on food that contains no nitrogen. A dog, or horse, or other animal, will starve if fed on starch alone. There must be a proper mixture of both nitrogenous and non-nitrogenous food, to build up and sustain the body.

197. Young animals, in whose bodies a rapid formation of muscle is going on, require a great deal of nitrogenous food, and all animals require more non-nitrogenous, or carbonaceous food, in cold weather than when the weather is warm. In very cold countries, and during the winter in temperate climates, men will eat more fat meat than in summer. The Esquimaux, a people that live in the intensely cold regions of the frigid zone, drink oil and melted fat, which are consumed in their bodies like fuel in a stove or fireplace, and supply heat for the body.

198. Fat in animals, like starch, sugar, and oil in plants, contains no nitrogen, and, when an animal is not fed, this fat wastes away first—in other words, is consumed. If the animal be exposed to great cold without extra food it cannot fatten. The reason, therefore, that animals protected from the cold of winter fatten much faster than when exposed, is, that what would accumulate as fat is used in keeping the animal warm.

199. The earthy, or inorganic matter, in plants is as necessary for animal growth as the organic matter. The bony skeleton consists chiefly of calcium

phosphate, with a little calcium carbonate and other mineral substances derived from plants. While plants contain everything necessary for animal growth, some portions are richer in salts and nitrogenous material, and are therefore considered to be of more value, as the grain of wheat, corn, and oats. Foods differ greatly in value, as every farmer knows. Many experiments have been made to determine the feeding power of various kinds of food, and elaborate tables have been drawn up to express their comparative value. Every farmer practically constructs such a table for himself, at least, he sets a different value on different substances, and buys and sells accordingly.

200. In the following table, common hay is taken as the standard, and the numbers opposite each substance show how many pounds of each contain nourishment equivalent to ten pounds of hay:

	lbs.		lbs.
Common hay ., ...	10	Turnips..........	50
Clover hay........	8 to 10	Cabbage	20 to 30
Green clover..... .	45 to 50	Peas and beans ...	3 to 5
Wheat-straw......	40 to 50	Wheat..........	5 to 6
Oat-straw........	20 to 40	Oats....	4 to 7
Pea-straw	10 to 15	Corn............	5
Potatoes	20	Oil-cake (linseed)..	2 to 4

Of course, such tables represent only **general** results. Much depends on the **quality** of the food, the form in which it is given, the condition of the animal to which it is fed, and other circumstances which the intelligent farmer understands.

APPENDIX.

THE following simple directions are given for the benefit of those who have no experience in science-teaching:

The exhibition of specimens of soils, plants, fertilizers, etc., and the performance of even a few experiments, will awaken in the minds of pupils a lively interest in the subject, cultivate the power of observation, and render the work of teaching pleasant and practical. Many illustrations, in addition to those mentioned, will doubtless suggest themselves to the minds of teachers.

The figures refer to corresponding numbers in the text:

(1.) Show to the class a specimen of soil, place a little on the end of a table-knife or spatula, and heat it over an alcohol-lamp until the organic matter is burned off.

(2.) Exhibit to the class a sample of *alluvial* soil from a creek or river bottom.

(3.) Exhibit pieces of sulphur, carbon, phosphorus, iron, lead, silver, etc., to illustrate what is meant by an *element*.

(4.) Pulverize in a mortar a small quantity of potassium chlorate, and mix with it about one-fourth its weight of manganese dioxide (black oxide of manganese). Place the mixture in a test-tube or small glass flask, and apply the heat of an alcohol-lamp. Oxygen will be disengaged, as may be shown by lowering into the flask or test-tube a lighted splinter. The flame will be greatly increased, and, if blown out, and the splinter be again thrust into the flask, provided a spark be left, it will burst into flame. If a piece of red-hot charcoal or burning sulphur be lowered into the flask, it will burn with great energy.

(5.) Place a few scraps of zinc in a deep wineglass, and pour over them a little water. Now pour in a little hydrochloric or sulphuric acid, and bubbles of hydrogen will rise through the liquid. Bring a lighted match to the mouth of the wineglass, and a slight explosion will ensue, caused by the union of the escaping hydrogen with oxygen of the air. By placing the zinc and acid in a bottle supplied with a cork and glass tube, the hydrogen may be burned as it issues from the tube, or may be collected in a receiver over a pneumatic cistern. Be very careful not to light the escaping hydrogen until all the air has been expelled from the bottle, as a mixture of hydrogen with oxygen contained in air unites with an explosion on application of flame. It is always best to surround the bottle with a towel or handkerchief before lighting the hydrogen. An explosion can then produce no bad effects.

(6.) Remove the cork, and thrust a lighted splinter into the bottle ; a slight explosion will occur, the splinter will be extinguished, and again lighted as it is taken out, while the hydrogen will continue to burn at the mouth of the bottle. A small piece of burning candle lowered into the bottle will show these results to better advantage.

(7.) Hold a tumbler or other glass vessel over the flame of burning hydrogen, and the water produced will condense on the cool surface.

(8.) Dissolve some sugar or salt in a tumbler of water.

(9.) Float a piece of cork on the surface of water, and place on it a small pellet of phosphorus. Now light the phosphorus, and cover it quickly with a glass receiver or wide-mouthed bottle, the mouth dipping under the surface of the water. After the phosphorus has ceased burning, the water will absorb the white fumes of the phosphorus oxide, and rise in the bottle. The gas which remains is nearly pure nitrogen. A lighted splinter or candle if brought into the nitrogen will be extinguished.

(10.) Rub together in a mortar a little ammonium chloride (sal-ammoniac) and quicklime. Ammonia will be given off, which can be detected by its pungent odor. Bring a piece of red litmus-paper in contact with this gas, and it will turn blue immediately.

(11.) Place a few bits of copper in a wineglass, and pour over them a little nitric acid. Red fumes of nitrogen tetroxide will be formed. The blue liquid that is left contains copper nitrate.

(12.) Dip a piece of paper in spirits of turpentine and light it. The black smoke consists of finely divided carbon. Hold a piece of window-glass over the flame of a candle, pushing it down upon the flame, and it will soon be coated with carbon.

(13.) Place some small pieces of marble or limestone in a deep wineglass or bottle, and pour over them a little hydrochloric acid. The brisk effervescence that takes place is caused by the escape of carbon dioxide. A burning splinter or candle will be extinguished if lowered into the bottle.

(14.) Burn a small piece of sulphur, and hold a red rose or other flower over the fumes. The flower will be bleached. A similar effect may be produced with a sulphur-match.

(15.) Place a splinter of wood in strong sulphuric acid, and it will turn black. The acid does not act on the carbon of the wood. Pour a little of the strong acid into a wineglass of water, and the water will become very hot, caused by its union with the acid. Caution should be observed in pouring this acid into water, and care be taken *not to pour the water into the acid.*

(16.) Exhibit a piece of phosphorus, and show how easily it can be set on fire.

(17.) Rub a match, and show how the friction causes it to inflame.

(18.) Place a little phosphorus on a dry plate, set it on fire, and quickly place over it a receiver, or wide-mouthed bottle. The white fumes that rise are phosphorus pentoxide. Water unites with this oxide to form phosphoric acid.

(19.) Place some manganese dioxide (black oxide of manganese) in a glass flask or test-tube, pour over it some strong hydrochloric acid, and apply a gentle heat. Chlorine, a yellowish-green gas, will be disengaged. By means of a properly bent glass tube reaching to the bottom of a bottle, this gas may be collected by displacement. Immerse in it a lighted taper or splinter, and the light will be extinguished. A red rose moistened and placed in a bottle of chlorine will be bleached in a short time.

(20.) Heat in a test-tube, or small flask, a little potassium iodide, manganese dioxide, and sulphuric acid. The beautiful violet-colored vapor of iodine will be disengaged. Substitute potassium bromide for the potassium iodide, and the deep-red vapor of bromine will be given off.

(21.) Drop a small piece of potassium on the surface of water. The metal will burn with a violet flame. Place a small piece of potassium on the wick of an alcohol-lamp, touch it with a piece of ice, and the lamp will be lighted.

(22.) Drop a piece of sodium on the surface of water. It will move around, decomposing the water with the disengagement of hydrogen. Place a piece in a few drops of water sprinkled on a board, and the sodium will burst into flame of a deep-yellow color.

(23.) Color some water blue with litmus. Add a drop of acid to turn it red, and then drop a piece of potassium or sodium on the surface. The blue color will be restored as the metal is consumed.

(24.) Heat in the flame of an alcohol-lamp a piece of magnesium ribbon, and show the brilliancy of its combustion.

(25.) Collect samples of different varieties of soils, and exhibit them to the class.

(26.) Heat some green leaves or grass on a spatula, metallic plate, or small shovel, until nothing is left but the ash.

(27.) Exhibit specimens of sugar, starch, albumen, or white of egg, etc.

(28.) Pour some clear lime-water into a wineglass, and by means of a straw or glass tube breathe through the liquid. The milky appearance is caused by the carbon dioxide from the lungs combining with the lime, and forming insoluble calcium carbonate.

(29.) Collect, and show to the class, a few plants with root, stem, and leaves.

(30.) Pour some water on a lump of unslaked lime. In a few minutes the mass will become hot, combine with a portion of the water, and crumble into powder.

(31.) Pour some hydrochloric acid on marl, and effervescence will take place, caused by the escape of carbon dioxide.

(32.) Mix a tablespoonful of plaster of Paris with a little water, and show how it sets, or hardens.

Exhibit to the class specimens of as many different fertilizers as can be obtained.

QUESTIONS.

THE following questions are intended merely to direct attention to the more important points in the text. Intelligent teachers will not, of course, confine themselves to any set form of questions, but secure as far as possible a mastery of the subject.

CHAPTER I.

THE DEVELOPMENT OF SCIENTIFIC AGRICULTURE.

1. What is agriculture? As an art, what does it teach? What, as a science?

2. How does it rank as an industrial pursuit? Why has it always been first in importance?

3. Why was less cultivation of the soil required in the early ages of the world? Where did men obtain food and clothing? What effect had the increase of population?

4. What is said of the progress of agriculture as a science? Why do some men say that *practical* and *scientific* farming are different?

5. Why has the progress of agriculture as a science been slow? What great men are mentioned as having been farmers or having written on farming? What is necessary to accomplish great results in any science?

6. What other reason is given for the slow progress of agriculture as a science? What is botany? Why should a farmer know something of it?

7. What is zoölogy? Why should the farmer know something of this science? What does geology treat of? Why should the farmer study it? With what does the mechanic supply the farmer? What does physics treat of? Why should the farmer have some knowledge of chemistry?

8. Why should the farmer have a general acquaintance with the sciences mentioned? What relation do they sustain to agriculture?

9. What sciences are of most importance to the farmer? What improvements in farming implements are mentioned?

10. What benefits has agriculture derived from chemistry?

11. How does the chemist proceed in his investigations? What is it proposed to show?

CHAPTER II.

THE ORIGIN, COMPOSITION, AND CLASSIFICATION OF SOILS.

12. What is the soil? Of what does it consist? Where does the organic matter come from? When burned, what is left? What do geologists suppose the earth to have once been? What happened as it cooled down? Of what does soil consist?

13. What is meant by mechanical agencies? What by chemical? Give examples? Are these agencies always at work?

14. What is included under the term "rock"? What agencies disintegrate rocks? How does water act? How does thorough cultivation help this disintegrating action?

15. With what do the elements of a soil vary? Does the soil always rest on the rock from which it was formed? What is an *alluvial* soil?

16. What causes the difference in the quality of soils? Are there many simple substances?

17. How many elements are certainly known? Which are gases? Liquids? Solids? Are there any permanent gases?

18. How many elements are concerned in the growth of

plants? Name the non-metallic elements. The metals. What other elements are sometimes present?

19. What is said of oxygen? How prepared? What should be mixed with the potassium chlorate?

20. Properties of oxygen? What is combustion? Give examples.

21. With what elements does oxygen unite? What are these compounds called? How named? What is potash? What is lime?

22. Give cases of oxidation without light and intense heat? What happens in breathing? Can animals live without oxygen? Can plants?

23. What is said of hydrogen? How is it prepared?

24. Properties of hydrogen? Result of its combustion?

25. How is water formed? What does it form when condensed in the air? Where do natural waters get the mineral matter contained in them? What kind of water is the purest? Why?

26. Where is nitrogen found? How much is in the air? How is it prepared?

27. Properties of nitrogen? Is it poisonous? How can oxygen, hydrogen, and nitrogen, be distinguished?

28. What is ammonia? How made? How much does water absorb? What is hartshorn? Why is ammonia called an alkali?

29. How is ammonia formed naturally? How does it get into plants?

30. What is nitric acid? What does it form with metals? What is nitre? Chili saltpetre? How used?

31. In what forms does carbon exist? Is it abundant in plants? What per cent. in sugar? In turpentine? What makes smoke black?

32. What is carbon dioxide? Where found? What are its properties?

33. What is silicon? Is it necessary to plants? Where is it found?

34. What is sulphur? What does it form in burning? Its use? What is sulphuric acid? What are its compounds called? Give examples.

35. What is phosphorus? Where found? Why must it be kept under water? Its use? How does a match burn?

36. What is phosphoric acid? What is calcium phosphate? What does it form? Why must soil contain phosphorus? How can it be supplied?

37. What is chlorine? What is hydrochloric acid? How is chlorine separated? Its properties? Uses? Where found?

38. What is iodine? Bromine? To what are they similar? Properties of fluorine? .Where found?

39. What is potassium? How kept? Why? What effect has water on it? What is caustic potash? What do all acids contain? How are salts of potassium formed? Where found?

40. Describe sodium. How does it act on water? What is common salt? Where found? In what respects are caustic potash and soda similar to ammonia? Their use?

41. Describe calcium? Lime? What are marble, lime-stone, and chalk? How changed to lime? What is gypsum?

42. Describe magnesium and aluminum. Where found? What are its properties?

43. Ores of iron? For what purpose used? Where is iron found? What is said about manganese?

44. How is the composition of a soil generally given? Is it easy to analyze a soil?

45. What constituent exists in the largest proportion in the analysis given? What in the smallest?

46. How much organic matter is found in soils? What elements are likely to be deficient?

47. How may soils be classified?

48. What are calcareous soils? Peaty? Heavy? Light?

49. How deep is the soil generally? What is the sub-soil? How may the depth of soil be increased? Its advantages?

CHAPTER III.

THE COMPOSITION OF PLANTS.

50. How many elements are found in plants? What are they?

51. What element in soils is not in plants? What in plants and not in animals?

52. What effect has heat on plants? How much water in turnips, cabbages, and potatoes? In cured hay? What is ash?

53. What four elements are consumed when a plant is burned? Why called organic? What becomes of them?

54. What becomes of phosphorus and sulphur? Chlorine and silicon? How much ash in wood? In tobacco? Corn? Which has in it the most phosphorus oxide, the ash of corn or potatoes? Chlorine? Potassium oxide?

55. What are ultimate elements? Proximate? Name some.

56. What are amylaceous and saccharine substances? Of what elements are they composed? Where is cellulose found? How converted into sugar?

57. What are pectose substances? Where found?

58. Name some vegetable acids. What elements in them?

59. Where are fats and oils found? How do they differ from sugar? What else are included in this group?

60. How do the albuminoid or protein groups differ from those mentioned? Where is pure albumen found? What do botanists mean by "albumen"? Where is gluten found? Vegetable casein?

61. How many elements are in these compounds? What has a knowledge of their chemical constitution done?

62. How do starch and sugar differ? Where is starch found?

63. How does grape-sugar differ from starch? Where does the change of one into the other take place? How can cotton be changed into sugar?

64. What takes place in the leaf?

65. Which crop mentioned in the table has the most starch? Albuminoids? Fat? Woody fibre? Ash?

66. Are the ash constituents important? What becomes of the products of combustion? What does a plant carry from the soil? What is the ultimate effect?

67. What does Liebig mean by a system of spoliation? What elements are scarcely ever deficient? In what condition may they exist? How can this be remedied?

68. What elements are likely to be deficient? Can the soil produce when any of the essential elements of plants are absent?

69. What does chemistry teach the farmer? For what must it not be held responsible?

70. By what must the farmer be guided?

CHAPTER IV.

THE COMPOSITION AND PROPERTIES OF THE ATMOSPHERE.

71. What does the atmosphere contain?

72. Is the air uniform in composition? In what condition are its constituents?

73. What is said of watery vapor? Whence does it rise? What does it form when condensed?

74. The character of nitrogen? Of oxygen? The effect of carbon dioxide on animals? On plants?

75. How is carbon dioxide formed? What prevents its accumulation in the atmosphere? What change takes place in it when absorbed by the leaves of plants?

76. How does carbon dioxide get into the leaves? Through what influence is it decomposed?

77. How much carbon dioxide exists in the atmosphere? What says a distinguished chemist?

78. How much ammonia is in the atmosphere?

79. To what is the atmosphere adapted? What benefits are derived from it?

80. What does science teach in reference to the atmospheric agencies? To what do they contribute? The effect of their action on solid substances?

81. The effect of the atmosphere on organic and inorganic matter? On a dead animal or plant?

82. What knowledge is necessary to understand the growth and development of plants?

CHAPTER V.

THE SOURCES OF PLANT-FOOD AND HOW OBTAINED.

83. The sources of plant-food? What elements are found in air?

84. Why cannot carbon enter the plant in a pure state? In what form is it a gas? How does it enter? Why do plants grow faster in daytime than at night? Where does the carbon of plants come from?

85. How do hydrogen and oxygen get into plants?

86. How nitrogen? Does free nitrogen contribute to growth? Where is it gotten?

87. How much nitrogen do plants contain? How much is in a ton of hay? Why must it be supplied in the form of salts containing nitrogen?

88. Why are the four elements mentioned called organic? What are the other constituents called? How do they get into plants? In what form?

89. What are the main parts of a plant? What does the seed contain? What does it require for germination? How does it germinate?

90. How does the root grow? What may be regarded as the lungs of a plant? What the mouth? What is said of the roots spreading? What is necessary for a plant to be thrifty? Why must its food be near at hand?

91. How did Schubert determine the extent of the roots of plants? How deep did he find the roots of wheat to extend? What is the total length of the roots of a barley-plant?

92. What increases the absorbing surface of roots?   Are there such organs as "spongioles"?

93. How many kinds of roots are there?   What is said of the mulberry-tree?   What of cornstalks?

94. Have roots the power of "excretion"?   Of selection?

95. What is evident from what has been said about roots? By what should the farmer be guided?

CHAPTER VI.

THE IMPROVEMENT OF SOILS.

96. What power has inert matter?   What has the seed? Do we know what the life-principle is?

97. Do we understand the connection between the growth of a plant and its surroundings?   What has chemistry taught?

98. What is necessary to insure fertility?   What vague notions have some men?

99. What have some writers asserted?   Is it true?   What can we not control?

100. What has science done?

101. What of sewage?   What does a perfect system of agriculture require?

102. Upon what are all scientific methods founded?   To what addressed?   What soil cannot be improved?   What can be?

103. What means can be used for the improvement of soils? What are mechanical?   What chemical?   The effect of the first?   Of the second?

104. What system of drainage is the cheapest?

105. What are the benefits of drainage?

106. What are the advantages of deep plowing and subsoiling?

107. What additional advantage is mentioned?   What care should be taken?

108. Mention an economical and effective method of subsoiling.

109. What other mechanical means of improvement are mentioned? What is the great object to be accomplished? What feeling generally prevails in a new country?

110. Is thorough cultivation always proper?

CHAPTER VII.

THE USE OF MANURES, OR FERTILIZERS.

111. Upon what principles is the use of fertilizers based?

112. When should fertilizers be used?

113. What produces exhaustion? How much potash is removed in a ton of red-clover hay? How much phosphoric acid?

114. How much phosphoric acid is contained in an ox weighing 1,000 pounds? How much potash in 1,000 pounds of unwashed wool?

115. Upon what does the rapidity of exhaustion depend? Is it sure to follow without restoration?

116. How many kinds of fertilizers are there?

117. Why is a decaying plant a good fertilizer? How do vegetable substances benefit land?

118. How do cornstalks serve as a fertilizer?

119. What is said of cotton-seed? Leaves?

120. What two important constituents are found in cotton-seed? How much mineral matter in a bale of lint? In cotton-seed from a bale?

121. If the seed be returned to the soil, is cotton an exhaustive crop? What escapes from fermenting seed? Why should it be preserved?

122. How should seed be applied?

123. Is the oil in cotton-seed valuable as a fertilizer? Why not?

124. What are the good effects of grinding the seed? How can its constituents be preserved?

125. What important object should be accomplished on every farm?

126. What are animal fertilizers ?   Are they valuable ?

127. What is said of the bones and excrements of animals as fertilizers ?

128. What is the composition of bones ?

129. How are they sometimes used ?

130. How are they converted into superphosphate ?

131. What does sulphuric acid form with calcium ?

132. Where has a large deposit of bone phosphate been found ?

133. What other source of superphosphate is mentioned ?

134. How can a good superphosphate be made by the farmer ?

135. Give the directions of Dr. Nichols for making it.

136. What is said of guano and its use in South America ?

137. What is guano ?

138. When was it introduced into Europe ?

139. What does Liebig say of its use ?

140. When was it brought to the United States ?

141. Why is the Peruvian guano the best ?

142. What makes guano so rich a fertilizer ?

143. Give the composition of guano.

144. How does Peruvian guano differ from the African ?

145. Give its properties.

146. To what is its stimulating effect due ?

147. What retards the growth of plants ?   Is nitrogen supplied from the air ?

148. How does guano act ?   Why does it sometimes " run out " ?

149. With what sort of fertilizers is it generally mixed ?

150. Why should it not be placed next to the seed ?

151. What is mentioned as a good mixture ?   How should it be applied ?

152. Should guano be mixed with lime ?   Why not ?

153. What is said of farm-yard manure ?

154. Upon what does its value depend ?   How can the escape of ammonia be determined ?   How prevented ?

155. How should manure-heaps be managed? What should be added?

156. What is said of human excrements?

157. Why are animal manures richer than vegetable?

CHAPTER VIII.

MINERAL FERTILIZERS.

158. What mineral fertilizers are mentioned?

159. How is lime made?

160. What is slaked lime? What effect on it has long exposure to air?

161. In what condition is the water in slaked lime? Its properties? How does mortar harden?

162. What is the use of lime as a fertilizer?

163. What is marl? How valued?

164. What is said of green-sand marl? Composition of marl?

165. What sort of lands are well suited for marl?

166. What is gypsum? What effect has heat upon it? How does it harden? Its use?

167. How found? Its properties? How applied?

168. Why does it not injure seeds? Why should it not be mixed with superphosphate?

169. What is common salt? How does it act?

170. What are salts? What is said of the utilization of waste products?

171. How much potash in the ashes of the oak? Beech?

172. Why are ashes a good fertilizer?

173. What is a compost? How made?

174. What directions are given by Pendleton for composting with cotton-seed?

175. How can a good compost be made from materials on every farm?

176. What does Professor Ville call a "complete manure"?

CHAPTER IX.

ROTATION OF CROPS.

177. What is meant by rotation ?

178. What are its advantages ? Give illustrations.

179. What other advantage is mentioned ?

180. Give a third advantage of rotation.

181. Give a fourth advantage.

182. What is said of change of seed ?

183. What crops should not follow each other ? What principle should govern ?

184. Mention a good three years' rotation. A four years' rotation. An English rotation.

185. A rotation for the tobacco-planter. For cotton-planter.

186 What is best to be done with the clover ? Why ?

187. Have plants the power of secretion ? Of selection ?

188. When will rotation certainly fail? How can such land be restored ?

CHAPTER X.

THE SELECTION AND CARE OF LIVE-STOCK.

189. Why must every farmer have live-stock ?

190. Upon what will the kind of stock to be kept depend ? What sort should be selected ? Why ?

191. What is necessary in the management of live-stock? Why? How should it be fed? Why should exposure be avoided ?

192. Difference between the functions of plants and animals ?

193. How do animals digest their food and convert it into bone and muscle ?

194. What gas is formed in respiration? How shown ?

195. Why should animals require more food in cold than in warm weather ? Why more when at work ?

196. What is Liebig's division of food ? Is his theory strict-ly true ?

197. What sort of food do young animals need ?

198. What first wastes away when an animal is not fed? Why do animals protected from cold fatten faster ?

199. What is the chief constituent of bones? Whence derived ?

200. IIow does wheat-straw compare with hay as a food for stock ? Upon what do the good effects of food depend ?

THE END.

Aids to Field and Laboratory Work in Botany

APGARS' PLANT ANALYSIS . . . . . 55 cents

A book of blank schedules, adapted to Gray's Botanies, for pupils' use in writing and preserving brief systematic descriptions of the plants analyzed by them in field or class work.   Space is allowed for descriptions of about one hundred and twenty-four plants with an alphabetical index.

An analytical arrangement of botanical terms is provided, in which the words defined are illustrated by small wood cuts, which show at a glance the characteristics named in the definition.

APGAR'S TREES OF THE NORTHERN UNITED STATES

Their Study, Description, and Determination   . $1.00

This work has been prepared as an accessory to the study of Botany, and to assist and encourage teachers in introducing into their classes instruction in Nature Study.   The trees of our forests, lawns, yards, orchards, streets, borders and parks afford a most favorable and fruitful field for the purposes of such study.   They are real objects of nature, easily accessible, and of such a character as to admit of being studied at all seasons and in all localities.   Besides, the subject is one of general and increasing interest, and one that can be taught successfully by those who have had no regular scientific training.

Copies will be sent, prepaid, on receipt of the price.

American Book Company

NEW YORK   •   CINCINNATI   •   CHICAGO
(S. 172)

Text-Books in Geology

By JAMES D. DANA, LL.D.

Late Professor of Geology and Mineralogy in Yale University

DANA'S GEOLOGICAL STORY BRIEFLY TOLD . $1.15

A new and revised edition of this popular text-book for beginners in the study, and for the general reader. The book has been entirely rewritten, and improved by the addition of many new illustrations and interesting descriptions of the latest phases and discoveries of the science. In contents and dress it is an attractive volume and well suited for young students.

DANA'S REVISED TEXT-BOOK OF GEOLOGY . $1.40

Fifth Edition, Revised and Enlarged. Edited by WILLIAM NORTH RICE, Ph.D., LL.D., Professor of Geology in Wesleyan University. This is the standard text-book in geology for high school and elementary college work. While the general and distinctive features of the original work have been preserved, the book has been thoroughly revised, enlarged, and improved.

DANA'S MANUAL OF GEOLOGY $5.00

Fourth Revised Edition. This great work is a complete thesaurus of the principles, methods, and details of the science of geology in its varied branches, including the formation and metamorphism of rocks, physiography, orogeny, and epeirogeny, biologic evolution, and paleontology. It is not only a text-book for the college student but a handbook for the professional geologist. This last revision was performed almost exclusively by Dr. Dana himself, and may justly be regarded as the crowning work of a long and useful life.

Copies sent, prepaid, on receipt of the price.

American Book Company

NEW YORK • CINCINNATI • CHICAGO

(S. 177)

A New Astronomy

By DAVID P. TODD, M.A., Ph.D.

Professor of Astronomy and Director of the Observatory, Amherst College

Cloth, 12mo, 480 pages. Illustrated. Price $1.30

This book is designed for classes pursuing the study of Astronomy in High Schools, Academies, and Colleges. The author's long experience as a director in astronomical observatories and in teaching the subject has given him unusual qualifications and advantages for preparing an ideal text-book.

The noteworthy feature which distinguishes this from other text-books on Astronomy is the practical way in which the subjects treated are reënforced by laboratory experiments and methods. In this the author follows the principle that Astronomy is preëminently a science of observation and should be so taught.

By placing more importance on the physical than on the mathematical facts of Astronomy the author has made every page of the book deeply interesting to the student and the general reader. The treatment of the planets and other heavenly bodies and of the law of universal gravitation is unusually full, clear, and illuminative. The marvelous discoveries of Astronomy in recent years, and the latest advances in methods of teaching the science, are all represented.

The illustrations are an important feature of the book. Many of them are so ingeniously devised that they explain at a glance what pages of mere description could not make clear.

Copies sent, prepaid, on receipt of the price.

American Book Company

NEW YORK   •   CINCINNATI   •   CHICAGO

(S. 181)

Text-Books in Chemistry

STORER AND LINDSAY'S ELEMENTARY MANUAL OF CHEMISTRY $1.20

This text-book is a thorough revision of Eliot, Storer, and Nichol's Elementary Manual of Chemistry, rewritten and enlarged to represent the present condition of chemical knowledge, and to meet the demands for a class book on Chemistry, for use in high schools or college preparatory schools.

CLARKE'S ELEMENTS OF CHEMISTRY . . . $1.20

A scientific book for high schools and colleges intended to provide a complete course for schools and to serve as a substantial basis for further study.

COOLEY'S NEW TEXT-BOOK OF CHEMISTRY . 90 cents

An elementary course designed for use in high schools and academies. The fundamental facts and principles are treated in a simple, concise, and accurate manner.

STEELE'S POPULAR CHEMISTRY $1.00

A popular treatise for high schools, academies, and private students.

BREWSTER'S FIRST BOOK OF CHEMISTRY . 66 cents

Designed to serve as a guide for beginners in the simplest elements of the science. The experiments are of the most elementary character, and only the simplest apparatus is employed.

LABORATORY METHODS

ARMSTRONG AND NORTON'S LABORATORY MANUAL OF CHEMISTRY . . . . 50 cents

COOLEY'S LABORATORY STUDIES IN CHEMISTRY . 50 cents

KEISER'S LABORATORY WORK IN CHEMISTRY 50 cents

IRISH'S QUALITATIVE ANALYSIS FOR SECONDARY SCHOOLS . . . . 50 cents

Copies will be sent, prepaid, on receipt of the price.

American Book Company

NEW YORK • CINCINNATI • CHICAGO

(S. 160)